Application of Geometric Algebra to Electromagnetic Scattering

Andrew Seagar

Application of Geometric Algebra to Electromagnetic Scattering

The Clifford-Cauchy-Dirac Technique

Springer

Andrew Seagar
Gold Coast, QLD
Australia

ISBN 978-981-10-0088-1 ISBN 978-981-10-0089-8 (eBook)
DOI 10.1007/978-981-10-0089-8

Library of Congress Control Number: 2015954612

Springer Singapore Heidelberg New York Dordrecht London

Printed on acid-free paper

Springer Science+Business Media Singapore Pte Ltd. is part of Springer Science+Business Media
(www.springer.com)

To Cathy

Preface

Electromagnetism immediately captured my interest when I was first introduced to it as a student at university. Complete mastery of one of the fundamental phenomena of our physical world seemed within reach, as promised in the form of Maxwell's equations [1][1]. However, the realisation of that promise was not as straightforward as I might have first imagined.

Exposure to electromagnetic theory in the period during and immediately following university, with its application to impedance imaging, left me with some skills—largely restricted to two dimensions and quasi-static fields. I found the conventional treatments at that time for fully three-dimensional time-varying fields clumsy and unpalatable. I took that to reflect my own inadequacy in grasping the methods rather than reflecting any inadequacy in the methods themselves. As it turns out, I was mistaken.

The first inkling that there might really be a simpler way came for me in 2002 with the paper of Axelsson, Grognard, Hogan and McIntosh [2]. At that time I started to seriously follow the literature relating to the theory of monogenic functions expressed in Clifford's algebra [3, 4] based on his interpretation of Hamilton's quaternions [5] in terms of Grassmann's extension theory [6, 7], and applied through the Cauchy integral [8, 9] to provide a solution to Maxwell's equations, and thereby to electromagnetic problems.

Although much is written, there are only a few works which I consider essential. First are the works of McIntosh [10, 11] providing a solid mathematical justification on the validity of the Cauchy integral in multiple dimensions for monogenic functions in Lipschitz domains. Second is the paper of Axelsson, Grognard, Hogan and McIntosh [2] giving the formulation of electromagnetic problems within the Clifford–Cauchy framework. Third is the thesis of Axelsson containing many more of the details [12].

Given that my background was in electrical engineering and the aforementioned documents were written by mathematicians, they probably would have remained

[1]Section 617.

forever beyond my understanding were it not that I had worked for many years with one of the authors. With his support and guidance I was slowly able to master the concepts and cast them, finally as this book, in a form by which they can be shared with others having a background at a level similar to my own.

The works in the book follow ultimately back to Cauchy, Hamilton, Grassmann, Maxwell and Clifford rather than conventionally to Green, Hamilton, Maxwell, Gibbs [13] and Heaviside [14][2] (also [15, 16]). The differences from conventional techniques are that (i) Hamilton's quaternions are used in whole to construct Clifford's algebra rather than being stripped apart to deliver Gibbs's vector calculus, (ii) Clifford's algebra is used instead of Gibbs's vector calculus and (iii) integrals are used involving Cauchy kernels instead of Green's functions.

The ultimate outcome is an integral formulation which lends itself to a more direct and efficient solution than is conventionally the case. That leads by necessity on any particular machine to either a faster solution for a given problem or the ability to solve problems of greater complexity. On the basis of efficiency alone, the new technique is destined in time to supplant at least some existing techniques.

The book is written with the intention that any who are interested can also come to master, over a much shorter time-span than in my case, techniques which offer simpler and more efficient solutions to problems of electromagnetic scattering than are currently in use.

During the development of the material and preparation of this work many people have been of assistance in various ways, either consciously or unknowingly, directly or indirectly. The various ways include the penetrating questions of students, the sound advice and encouragement of colleagues and peers, the perceptive counsel of elders and betters, the professional guidance of the editorial team, and the stoic forbearance and unfaltering faith of my immediate family.

Gold Coast
September 2015

Andrew Seagar

[2]Sections 51 and 66.

Contents

Part II Formulation

Part III Demonstration

Part IV Contemplation

Nomenclature

$\neg$	$\overset{\neg}{a}$	Involution	
$\sim$	$\tilde{a}$	Reversion	
$-$	$\bar{a}$	Clifford conjugation	
$\bigwedge$	$\bigwedge^a$	Grade	
$\cdot$	$a \cdot b$	Dot product	
$\lrcorner$	$a \lrcorner b$	Left inner product	
$\llcorner$	$a \llcorner b$	Right inner product	
$(,)$	(a, b)	Scalar product	
$\wedge$	$a \wedge b$	Wedge product, outer product	
$\vee$	$a \vee b$	Inner product	
$\cap$	$a \cap b$	Evolution, outer product, intersection of sets	
$\times$	$a \times b$	Cross product	
∂	∂a	Partial differential	Scalar
$\reflectbox{$\partial$}$	$a\reflectbox{$\partial$}$	Partial differential operating towards the left	Scalar
D	Da	Differential operator	Generic
∇	∇a	Gradient operator	Vector
$\nabla \cdot$	$\nabla \cdot a$	Divergence operator	Vector
$\nabla \times$	$\nabla \times a$	Curl operator	Vector
$\boldsymbol{\nabla}$	$\boldsymbol{\nabla} a$	Gradient operator	Clifford
$\mathcal{D}$	$\mathcal{D}a$	Hyperbolic Dirac operator	Clifford
$\mathcal{D}_0$	$\mathcal{D}_0 a$	Dirac operator	Clifford
$\mathcal{D}_k$	$\mathcal{D}_k a$	k-Dirac operator	Clifford
α, β		Magnetic and electric material contrast factors	Scalar
Γ		Reflection coefficient	Scalar
δ		Dirac delta function	Scalar
$\delta^{\pm}$		Point/dipole source	Clifford
$\epsilon, \epsilon_{\pm}$		Electric permittivity	Scalar
$\eta, \eta_0, \eta_{\pm}$		Intrinsic impedances	Scalar
λ		Wavelength	Scalar

(continued)

(continued)

$\mu, \mu_\pm$		Magnetic permeability	Scalar
$\rho; \rho_k$		Electric charge density; phasor	Scalar
ρ_s		Surface electric charge density	Scalar
σ		Negative of volume unit in three dimensions	Clifford
$\sigma_\pm$		Electrical conductivity	Scalar
$d\sigma$		Elemental measure of boundary	Scalar
Σ		Boundary or interface	Surface
τ		Transmission coefficient	Scalar
ϕ		Phase of travelling wave	Scalar
$\phi; \phi_k$		Electric potential; phasor	Scalar
ω		Angular frequency	Scalar
$\Omega, \Omega^\pm$		Regions or objects	Volume
$A; A_k$		Magnetic potential; phasor	Clifford
$\mathbf{A}$		Magnetic potential	Vector
$\mathcal{A}, \mathcal{B}, \mathcal{C}, \mathcal{D}, \mathcal{Q}$		Quaternions	Quaternion
$B_{\pm k}$		Bessel potential	Scalar
B, D		Magnetic and electric flux densities	Clifford
$\mathbf{B}$, $\mathbf{D}$		Magnetic and electric flux densities	Vector
c		Speed of light	Scalar
C		Cauchy extension	Clifford
$C_\Sigma, C_\Sigma^\pm$		Cauchy integrals	Clifford
$e_\emptyset$		Absolute unit	Scalar
e_0		Primal time/frequency unit	Clifford
$e_p, e_q, e_r, e_s \ldots$		Primal (vector) units	Clifford
e_1, e_2, e_3		Primal (vector) units in three dimensions	Clifford
E_k		Cauchy kernel	Clifford
$E, H; E_k, H_k$		Electric and magnetic fields; phasors	Clifford
$\mathbf{E}$, $\mathbf{H}$		Electric and magnetic fields	Vector
$\mathrm{E}_\star, \mathrm{H}_\star$		Incident fields	Vector
$\mathrm{E}_\pm, \mathrm{H}_\pm$		Scattered and transmitted fields	Vector
f		Field, joint interface function	Clifford
$f^\star$		Incident field	Clifford
$f^\pm$		Scattered and transmitted fields	Clifford
F		Function in Minkowskian space–time	Clifford
F_k		Fundamental solution	Clifford
$\mathcal{F}; \mathcal{F}_k$		Electromagnetic field; phasor	Clifford
g		Boundary data vector	Clifford
H_k		Helmholtz operator	Scalar
i		Imaginary unit	Scalar
I		Identity operator	
$I_{RQ}^\oplus, I_{RQ}^\ominus$		Images of axes	Line
$\mathbf{I}$, $\mathbf{J}$, $\mathbf{K}$		Quaternion units	Quaternion

(continued)

(continued)

$J; J_k$		Electric current density; phasor	Clifford
$\mathbf{J}$		Electric current density	Vector
$\mathbf{J}_s$		Surface electric current density	Vector
$k, k_\pm$		Wavenumber	Scalar
n		Unit normal vector	Clifford
$\mathbf{n}$		Unit normal vector	Vector
O		Origin of axes	Point
$OP, OP'; OQ, OQ'$		Pairs of coordinate axes	Line
$OP^\oplus, OP^\ominus$		Unrelated axes	Line
$P^\pm, P^{\pm\pm}$		Hardy projection operators	Clifford
$\mathcal{P}; \mathcal{P}_k$		Electromagnetic potential; phasor	Clifford
$Q^\pm$		Boundary data projection operators	Clifford
$R, R_{\alpha\beta}$		Space–time rescaling operator	Clifford
$\mathcal{S}; \mathcal{S}_k$		Electromagnetic source; phasor	Clifford
S, T		Space-like and time-like projection operators	Clifford
t		Time	Scalar
$U_{(m)}, V_{(m)}, W_{(m)}$		Clifford numbers of dimension m	Clifford
x_0, x_1, x_2, x_3		Coordinates in Minkowskian space–time	Real
X_0, X_1, X_2, X_3		Basis vectors in Minkowskian space–time	Clifford
$\mathbb{X}, \mathbb{Y}$		Inertial frames of reference	
$\hat{\mathbf{x}}, \hat{\mathbf{y}}, \hat{\mathbf{z}}$		Vector units in three dimensions	Vector
$X_{(m)}, Y_{(m)}, Z_{(m)}$		Clifford numbers of dimension m	Clifford
y_0, y_1, y_2, y_3		Coordinates in Minkowskian space–time	Real
Y_0, Y_1, Y_2, Y_3		Basis vectors in Minkowskian space–time	Clifford
$Z_\pm$		Characteristic impedances	Scalar

List of Figures

List of Tables

Part I
Preparation

Chapter 1
History

1.1 Grassmann's Outer Product

1.1.1 Evolution

Grassmann describes an abstract operation called *evolution*, which he illustrates by applying it in a particular form to generate higher-dimensional geometric objects from lower-dimensional ones [6]. Figure 1.1 shows the evolution of one vector **a** (called the *generating element*) as specified by another vector **b** (called the *measure of generation*). For this geometric application the operation of evolution is taken as the process of translation of the generating element **a** through a continuous sequence of intermediate states along a locus having the same length and direction as the measure of generation **b**. At each intermediate state the image of the generating element (the broken arrow) is called an *intermediate element* of the continuous form.

The result of the evolution is taken as the whole collection of intermediate elements into which the generating element is transformed during the operation. With an initial generating element **a** and a measure of generation **b**, the result is the interior of the parallelogram having sides coincident with the vectors **a** and **b**.

Grassmann's term for the result of his abstract concept of evolution is the *extension*. In the geometric application here the operation of evolution on a one-dimensional object produces an extension which is a two-dimensional object. The process can be repeated by taking the parallelogram as the generating element and subjecting that to the operation of evolution as specified by a third vector **c**. As shown in Fig. 1.2 the result (the extension) in this case is the interior of a parallelepiped having sides coincident with the vectors **a**, **b** and **c**.

The single application of evolution producing the parallelogram was originally written by Grassmann as $\mathbf{a} \cap \mathbf{b}$, and the double application of evolution producing the parallelepiped was written $\mathbf{a} \cap \mathbf{b} \cap \mathbf{c}$. In this notation it is understood implicitly that the operations are performed sequentially from left to right, so that $\mathbf{a} \cap \mathbf{b} \cap \mathbf{c} = (\mathbf{a} \cap \mathbf{b}) \cap \mathbf{c}$.

A. Seagar, *Application of Geometric Algebra to Electromagnetic Scattering*,
DOI 10.1007/978-981-10-0089-8_1

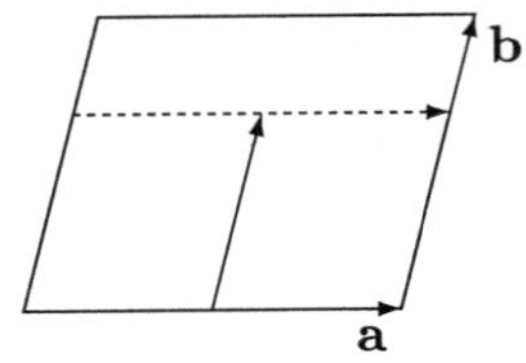

Fig. 1.1 Evolution of generating element **a**, as specified by measure of generation **b**

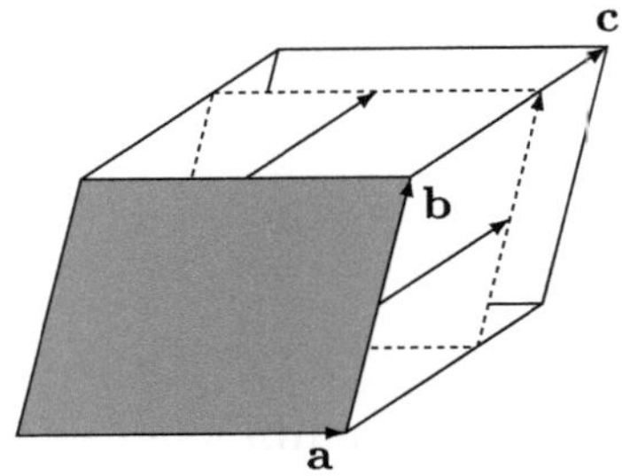

Fig. 1.2 Evolution of parallelogram, as specified by measure of generation **c**

Grassmann's definition of evolution for the geometric application essentially plays the role of integration, with the vector **a** taking the role of integrand and the vector **b** taking the role of the locus along which the integration is performed. In general for a generating element **a** and a differential measure of generation $\mathbf{b}'\,db$ the integral can be written $\int_{\mathbf{b}} \mathbf{a} \cap \mathbf{b}'\,db$. In Grassmann's more specific case where the vector functions **a** and $\mathbf{b}'$ are constant functions, the expression reduces to $\mathbf{a} \cap \int_{\mathbf{b}} \mathbf{b}'\,db = \mathbf{a} \cap \mathbf{b}$, where **b** is the vector from start to end of the locus.

1.1.2 Outer Product

When the generating element **a** and the measure of generation **b** are each composed by the addition of two linearly dependent parts, $\mathbf{a}_1$, $\mathbf{a}_2$ and $\mathbf{b}_1$, $\mathbf{b}_2$ respectively, then the evolution $\mathbf{a} \cap \mathbf{b} = (\mathbf{a}_1 + \mathbf{a}_2) \cap (\mathbf{b}_1 + \mathbf{b}_2)$ can be completed as shown in Fig. 1.3 in four parts: $(\mathbf{a}_1 \cap \mathbf{b}_1) + (\mathbf{a}_1 \cap \mathbf{b}_2) + (\mathbf{a}_2 \cap \mathbf{b}_1) + (\mathbf{a}_2 \cap \mathbf{b}_2)$.

Taking in turn **a** or **b** as whole gives the two relationships:

$$\begin{cases} \mathbf{a} \cap (\mathbf{b}_1 + \mathbf{b}_2) = (\mathbf{a} \cap \mathbf{b}_1) + (\mathbf{a} \cap \mathbf{b}_2) \\ (\mathbf{a}_1 + \mathbf{a}_2) \cap \mathbf{b} = (\mathbf{a}_1 \cap \mathbf{b}) + (\mathbf{a}_2 \cap \mathbf{b}) \end{cases} \tag{1.1}$$

Apart from the symbols chosen, the relationships given in Eq. 1.1 exhibit exactly the distributive behaviour expected of a multiplication-like operation $\cap$ over a addition-like operation $+$ from either left or right. It is therefore possible to identify from its behaviour the evolution $\cap$ as defined in Figs. 1.1 and 1.2 as some kind of geometric multiplication. Grassmann called this geometric multiplication the *outer* product. Today it is more common to write $\mathbf{a} \wedge \mathbf{b}$ instead of $\mathbf{a} \cap \mathbf{b}$ and call this multiplication

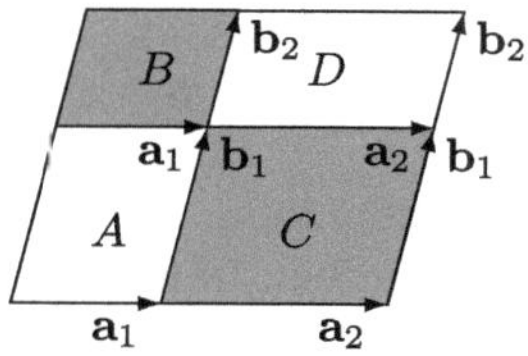

Fig. 1.3 Evolution of generating element $\mathbf{a}$ in two parts $\mathbf{a}_1$, $\mathbf{a}_2$, as specified by measure of generation $\mathbf{b}$ in two parts $\mathbf{b}_1$, $\mathbf{b}_2$. $A = \mathbf{a}_1 \cap \mathbf{b}_1$, $B = \mathbf{a}_1 \cap \mathbf{b}_2$, $C = \mathbf{a}_2 \cap \mathbf{b}_1$, $D = \mathbf{a}_2 \cap \mathbf{b}_2$

either the outer product or the *wedge* product. At other times Grassmann also used alternative notations for the outer product, namely $\mathbf{a} \cdot \mathbf{b}$ (which is now used for the dot product) and $[\mathbf{ab}]$.

1.1.3 Algebraic Properties

Further algebraic properties of the outer product can be determined from its geometric definition taken together with its distributive properties.

A. Single Products

Figure 1.4 shows through geometric and algebraic steps along the path indicated by double ended arrows that $\mathbf{a} \cap \mathbf{b} = \mathbf{a} \cap (\mathbf{a}_1 + \mathbf{b})$.

The product of two vectors which are linearly dependent is therefore $\mathbf{a} \cap \mathbf{a}_1 = 0$. Choosing $\mathbf{a}_1 = \mathbf{a}$ then gives $\mathbf{a} \cap \mathbf{a} = 0$ for any vector $\mathbf{a}$, so that $0 = (\mathbf{a} + \mathbf{b}) \cap (\mathbf{a} + \mathbf{b}) = (\mathbf{a} \cap \mathbf{a}) + (\mathbf{a} \cap \mathbf{b}) + (\mathbf{b} \cap \mathbf{a}) + (\mathbf{b} \cap \mathbf{b}) = (\mathbf{a} \cap \mathbf{b}) + (\mathbf{b} \cap \mathbf{a})$. This is possible only if $\mathbf{a} \cap \mathbf{b} = -\mathbf{b} \cap \mathbf{a}$, which indicates that the outer product is anti-commutative. Notice that the algebraic properties of the outer (wedge) product:

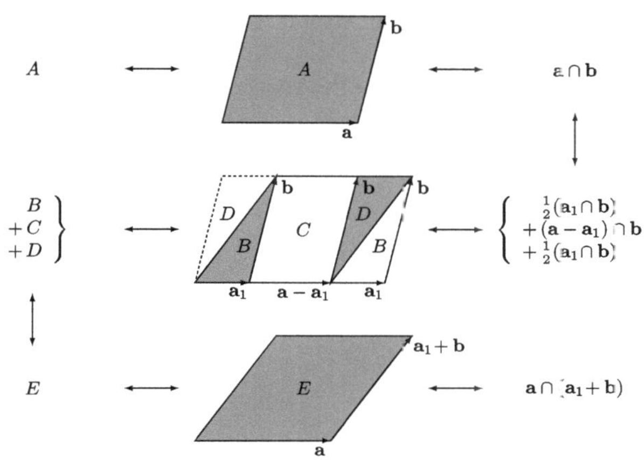

Fig. 1.4 Demonstration that $\mathbf{a} \cap \mathbf{b} = \mathbf{a} \cap (\mathbf{a}_1 + \mathbf{b})$ using chain of geometric and algebraic reasoning (*arrows*)

$$\begin{cases} \mathbf{a} \cap \mathbf{a} = 0 \\ \mathbf{a} \cap \mathbf{b} = -\mathbf{b} \cap \mathbf{a} \end{cases} \tag{1.2}$$

come not as an abstract algebraic definition, but by construction from the geometric operation it was intended to represent. In mathematical terminology, the vector operations of addition and outer product form a ring which is neither commutative nor unitary.

The anti-commutative property of the product means that geometrically the parallelogram $\mathbf{a} \cap \mathbf{b}$ and the parallelepiped $\mathbf{a} \cap \mathbf{b} \cap \mathbf{c}$ play the roles of *oriented areas* and *oriented volumes* respectively.

B. Double Products

The double product $\mathbf{a} \cap \mathbf{b} \cap \mathbf{c} = (\mathbf{a} \cap \mathbf{b}) \cap \mathbf{c}$ produces a parallelepiped having sides coincident with the three vectors **a**, **b** and **c**. Performing the same operation with the same vectors in a different order produces another parallelepiped with the same shape, although possibly with a different sign. For example, $(\mathbf{a} \cap \mathbf{b}) \cap \mathbf{c} = -(\mathbf{b} \cap \mathbf{a}) \cap \mathbf{c}$ because the outer product of vectors is anti-commutative. The sign plays the role of converting a left handed orientation into a right hand orientation.

Figure 1.5 shows the parallelepipeds produced using the same three vectors in different orders: $(\mathbf{a} \cap \mathbf{b}) \cap \mathbf{c}$, $(\mathbf{b} \cap \mathbf{c}) \cap \mathbf{a}$ and $(\mathbf{c} \cap \mathbf{a}) \cap \mathbf{b}$. The intermediate elements of the continuous form, produced from the generating elements as they evolve over the surfaces of the parallelepipeds, are indicated by broken arrows. The position of these intermediate elements relative to one another defines the orientation of the parallelepipeds. Note that this 'orientation' refers to the orientation of evolution (or construction) of a parallelepiped rather than its physical orientation in space.

Making a rotation of the parallelepipeds around the axis AA' to alter the physical orientation by some $\pm 120°$ aligns the pictorial representation of the evolutional orientation for all three. Observing that the alignment is possible confirms that, in addition to the shape, the sign (evolutional orientation) of the three parallelepipeds is also the same. Combining the two observations gives:

$$(\mathbf{a} \cap \mathbf{b}) \cap \mathbf{c} = (\mathbf{b} \cap \mathbf{c}) \cap \mathbf{a} = (\mathbf{c} \cap \mathbf{a}) \cap \mathbf{b} \tag{1.3}$$

The ability to interchange the vectors of the double product in this particular manner without altering the result implies that the algebra of the outer product adheres to

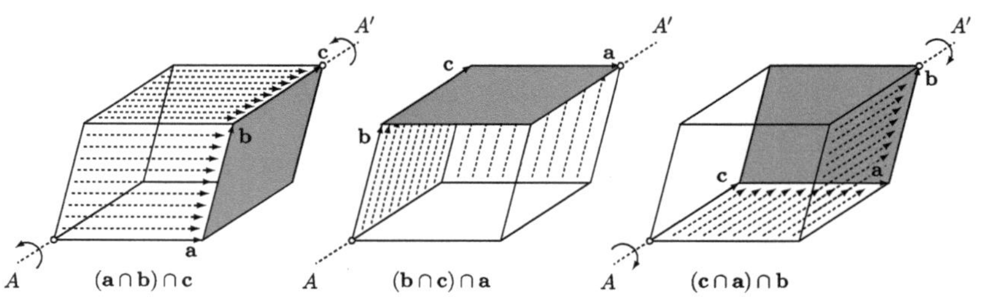

Fig. 1.5 Parallelepipeds produced from vectors **a**, **b**, **c** in different orders

some kind of associative property. Hypothesising that the associative property is of the form:

$$(\mathbf{a} \cap \mathbf{b}) \cap \mathbf{c} = \mathbf{a} \cap (\mathbf{b} \cap \mathbf{c}) \tag{1.4}$$

and applying that property together with the anti-commutative property gives:

$$(\mathbf{a} \cap \mathbf{b}) \cap \mathbf{c} \rightarrow \mathbf{a} \cap (\mathbf{b} \cap \mathbf{c}) \rightarrow -\mathbf{a} \cap (\mathbf{c} \cap \mathbf{b}) \rightarrow -(\mathbf{a} \cap \mathbf{c}) \cap \mathbf{b} \rightarrow (\mathbf{c} \cap \mathbf{a}) \cap \mathbf{b} \tag{1.5}$$

in agreement with observed geometrical behaviour. The hypothesis can therefore be allowed to stand.

The chain of reasoning which verifies the associative behaviour of the outer product (Eq. 1.5) involves terms of the form $\mathbf{a} \cap (\mathbf{b} \cap \mathbf{c})$. The evolution represented by these terms can be interpreted geometrically as a special case of a two-dimensional integral, $\int_{\mathrm{b}} \mathbf{a} \cap \int_{\mathrm{c}} \mathbf{b}' \cap \mathbf{c}' \, dc \, db$, in which the three vector functions $\mathbf{a}$, $\mathbf{b}'$ and $\mathbf{c}'$ are all constant functions. With this interpretation the measure of generation for the generating element $\mathbf{a}$ is seen as a two-dimensional oriented area $\mathbf{b} \cap \mathbf{c}$ rather than, as in the case of a single product $\mathbf{a} \cap \mathbf{b}$, a one-dimensional oriented path $\mathbf{b}$.

1.2 Three Kinds of Multiplication

1.2.1 Constraint Equations

Grassmann called the Eq. 1.2 which describe how vectors combine under outer multiplication the *constraint equations* for that particular kind of multiplication [17–20].[1] Other kinds of multiplication adhere to other constraint equations.

Grassmann identified different classes of multiplication based on different sets of constraint equations. Outer multiplication belongs to a class called *circular multiplication* [17, 19]. Different kinds of circular multiplication adhere to one or more of the constraint equations listed in Table 1.1, where the vector units e_p serve as a basis in terms of which any vector $\mathbf{a} = a_1e_1 + a_2e_2 + \cdots + a_me_m$ may be expressed.

The outer product is generated by imposing the second and third constraint equations as shown in Table 1.2. The *inner* product, symbolised for vectors by a dot in contemporary notation, is generated by using the first and second constraint equations. The product which Grassmann called *central* is generated by using the second constraint equation alone [18, 20].

In the case of the inner and central products there is an arbitrary scale factor which can be freely chosen. For the inner product the choice is plus one, and for the central product the choice is minus one. The latter choice was made by Grassmann in order to show that Hamilton's quaternions [5] could be constructed from that particular central product [18, 20]. Clifford later used the same normalisation, for much the same

[1]There are typographical errors in the English translations which can be identified by referring to the original documents.

Table 1.1 Constraint equations for circular multiplication

1	$e_r e_s = e_s e_r$
2	$e_r e_s + e_s e_r = 0$ and $e_1^2 = e_2^2 = \cdots = e_m^2$
3	$e_1^2 + e_2^2 + \cdots + e_m^2 = 0$

Table 1.2 Three circular multiplications

Product	Constraint equation	Unscaled	Scaled	Contemporary notation
Inner	1 and 2	$e_r e_s = 0$	$e_r e_s = 0$	$e_r \cdot e_s = 0$
		$e_r e_r = e_s e_s$	$e_r^2 = +1$	$e_r \cdot e_r = +1$
Central	2	$e_r e_s = -e_s e_r$	$e_r e_s = -e_s e_r$	$e_r e_s = -e_s e_r$
		$e_r e_r = e_s e_s$	$e_r^2 = -1$	$e_r^2 = -1$
Outer	2 and 3	$e_r e_s = -e_s e_r$	$e_r e_s = -e_s e_r$	$e_r \wedge e_s = -e_s \wedge e_r$
		$e_r e_r = 0$	$e_r^2 = 0$	$e_r \wedge e_r = 0$

reason in constructing an algebra [3] for the purpose of solving Maxwell's quaternion equations for electromagnetism [1].[2] In retrospect, the constraint equations for the central product and Clifford's algebra are the same.

The choice $e_r^2 = -1$ for the central product made by Grassmann and adopted by Clifford is not the only one followed. As a result, other conventions where $e_r^2 = +1$, or where $e_r^2 = +1$ for some values of r and zero or -1 for other values of r, are also used. Here, Grassmann's and Clifford's convention is followed.

1.2.2 Interdependence

Grassmann's three kinds of circular multiplication: inner, central and outer, are not independent. The central product can be expressed in terms of the other two:

$$\mathbf{ab} = \mathbf{a} \wedge \mathbf{b} - \mathbf{a} \cdot \mathbf{b} \tag{1.6}$$

The relationship is verified in Table 1.3 by substitution of the two cases $e_r = e_s$ and $e_r \neq e_s$ into Eq. 1.6 and use of the equations in Table 1.2.

The interdependence can be exploited to calculate the inner and outer products from the central product:

$$\begin{cases} \mathbf{a} \cdot \mathbf{b} = -\frac{1}{2}(\mathbf{ab} + \mathbf{ba}) \\ \mathbf{a} \wedge \mathbf{b} = +\frac{1}{2}(\mathbf{ab} - \mathbf{ba}) \end{cases} \tag{1.7}$$

[2]Section 617.

Table 1.3 Verification of relationship between inner, central and outer products

Case	Central	Outer	Inner	Outer–Inner
$e_r = e_s$	-1	0	1	-1
$e_r \neq e_s$	$e_r \wedge e_s$	$e_r \wedge e_s$	0	$e_r \wedge e_s$

1.3 Complex Numbers and Quaternions

1.3.1 Quaternions

Hamilton developed quaternions [5] as a generalisation of complex numbers. Quaternions Ω are constructed as a linear combination of the elements from a set of four quaternion units $\{\mathbf{1}, \mathbf{I}, \mathbf{J}, \mathbf{K}\}$ weighted by constant coefficients ξ, such as $\Omega = \xi_0\mathbf{1} + \xi_1\mathbf{I} + \xi_2\mathbf{J} + \xi_3\mathbf{K}$. The quaternion units adhere to the constraint equations:

$$\begin{cases} \mathbf{IJ} = \mathbf{K} \\ \mathbf{JK} = \mathbf{I} \\ \mathbf{KI} = \mathbf{J} \end{cases} \quad \text{and} \quad \mathbf{IJK} = -\mathbf{1} \tag{1.8}$$

Note that these quaternion constraint equations simultaneously accommodate the constraint equations $\mathbf{I}^2 = -\mathbf{1}$ and $\mathbf{IJ} = -\mathbf{JI}$ of Grassmann's central multiplication.

1.3.2 Complex Numbers

Complex numbers are retrieved from quaternions by setting the coefficients $\xi_3 = \xi_2 = 0$. This can be verified by multiplying two such numbers $u = v\mathbf{1} + w\mathbf{I}$ and $z = x\mathbf{1} + y\mathbf{I}$:

$$\begin{aligned} uz = (v\mathbf{1} + w\mathbf{I})(x\mathbf{1} + y\mathbf{I}) &= vx\mathbf{1} + wy\mathbf{I}^2 + vy\mathbf{I} + wx\mathbf{I} \\ &= (vx - wy)\mathbf{1} + (vy + wx)\mathbf{I} \end{aligned} \tag{1.9}$$

Noting that imaginary unit i follows the same constraint equation $i^2 = -1$ as does the quaternion unit $\mathbf{I}$, it then follows, if the quaternion units $\mathbf{1}$ and $\mathbf{I}$ are replaced by the real and imaginary units 1 and i respectively, that the exact same result represents the product of two complex numbers.

1.4 Clifford's Algebra

Clifford [3] constructed a set of quaternion units $\mathbf{I}_1, \mathbf{J}_1, \mathbf{K}_1$ from the vector units e_1, e_2, e_3 (or $\iota_1, \iota_2, \iota_3$ in his notation):

$$\begin{cases} \mathbf{I}_1 &= e_2 e_3 \\ \mathbf{J}_1 &= e_3 e_1 \\ \mathbf{K}_1 &= e_1 e_2 \end{cases} \tag{1.10}$$

with the multiplication being that of Grassmann's central product. Hamilton's rules for quaternion units (Eq. 1.8) then follow directly. With this construction the quaternion units can be interpreted as extensions of the second order, i.e. two-dimensional oriented areas. However the construction is not unique, so neither is its interpretation.

Clifford realised that it is also possible to construct a second set of quaternions, which adhere to Hamilton's rules for quaternion units and which exhibit commutativity under central multiplication with the original set but do not interact with them in any other way. Consider for example one such set:

$$\begin{cases} \mathbf{I}_2 &= \quad e_0 \\ \mathbf{J}_2 &= -i\sigma \\ \mathbf{K}_2 &= -ie_0\sigma \end{cases} \tag{1.11}$$

where e_0 is a fourth vector unit, i is the imaginary unit, and $\sigma = -e_1 e_2 e_3$ for notational convenience. This particular set is worthy of note because it conforms in a direct fashion to the vector calculus later developed by Gibbs [13]. Clifford himself made a different choice [3], there not being at that time any need to conform.

Clifford constructed elements C of his algebra as quaternions:

$$C = \mathfrak{A}\mathbf{1} + \mathfrak{B}\mathbf{I}_2 + \mathfrak{C}\mathbf{J}_2 + \mathfrak{D}\mathbf{K}_2 \tag{1.12}$$

with quaternion coefficients $\mathfrak{A}$, $\mathfrak{B}$, $\mathfrak{C}$, $\mathfrak{D}$ of the form $\mathfrak{A} = \xi_0 \mathbf{1} + \xi_1 \mathbf{I}_1 + \xi_2 \mathbf{J}_1 + \xi_3 \mathbf{K}_1$ for real or complex scalars $\xi_0 \ldots \xi_3$. Because the quaternion units from the two sets do not interact with one another, the algebra of these quaternion-quaternions has sixteen independent units formed by the pairwise product of all elements in the two sets:

$$\{\mathbf{1}, \mathbf{I}_1, \mathbf{J}_1, \mathbf{K}_1\} * \{\mathbf{1}, \mathbf{I}_2, \mathbf{J}_2, \mathbf{K}_2\} \tag{1.13}$$

Clifford called this algebra a *geometric algebra.* Today this geometric algebra is called a four-dimensional Clifford algebra.

1.5 Vector Calculus

The rules of Gibbs's three-dimensional vector calculus [13] are based on two vector products called the dot $(\cdot)$ and cross $(\times)$ products. These vector products can both be calculated from the central product:

$$\begin{cases} \mathbf{a} \cdot \mathbf{b} &= -\frac{1}{2}(\mathbf{ab} + \mathbf{ba}) \\ \mathbf{a} \times \mathbf{b} &= +\frac{1}{2}(\mathbf{ab} - \mathbf{ba})\sigma \end{cases} \tag{1.14}$$

where $\sigma = -e_1e_2e_3$. Comparison of Eq. 1.14 with Eq. 1.7 shows the dot product as identical to the inner product, and the cross product as closely related to the outer (wedge) product:

$$\mathbf{a} \times \mathbf{b} = (\mathbf{a} \wedge \mathbf{b})\sigma \tag{1.15}$$

Whereas the result of the outer product of two vectors is an oriented area, the result of the cross product is the oriented normal to that area.

In three dimensions the central product can be expressed in terms of dot and cross products:

$$\mathbf{ab} = (\mathbf{a} \times \mathbf{b})\sigma - \mathbf{a} \cdot \mathbf{b} \tag{1.16}$$

Putting $\mathbf{b} = \mathbf{a}$ shows the central product square of a vector is minus the square of its magnitude $a = |\mathbf{a}|$:

$$\mathbf{a}^2 = \mathbf{aa} = (\mathbf{a} \times \mathbf{a})\sigma - \mathbf{a} \cdot \mathbf{a} = -\mathbf{a} \cdot \mathbf{a} = -|\mathbf{a}|^2 = -a^2 \tag{1.17}$$

and the central product reciprocal of a vector is then:

$$\frac{1}{\mathbf{a}} = \frac{\mathbf{a}}{\mathbf{a}^2} = -\frac{\mathbf{a}}{|\mathbf{a}|^2} = -\frac{\mathbf{a}}{a^2} \tag{1.18}$$

1.6 Synchronicity

A four-dimensional Clifford algebraic value $\mathcal{C}$ can be expressed in terms of components which are either vector units (Grassmann), quaternions (Hamilton) or vectors (Gibbs). The correspondence between these alternative expressions is formulated as:

$$\begin{aligned} \mathcal{C} &= \mathfrak{A}\mathbf{1} + \mathfrak{B}\mathbf{I}_2 + \mathfrak{C}\mathbf{J}_2 + \mathfrak{D}\mathbf{K}_2 \\ &= \mathfrak{A} + \mathfrak{B}e_0 - i\mathfrak{C}\sigma - i\mathfrak{D}e_0\sigma \\ &= (a + A\sigma) + (b + B\sigma)e_0 - i(c + C\sigma)\sigma - i(d + D\sigma)e_0\sigma \end{aligned} \tag{1.19}$$

where $\mathfrak{A}$, $\mathfrak{B}$, $\mathfrak{C}$, $\mathfrak{D}$ are quaternions based on one set of quaternion units $\{\mathbf{1}, \mathbf{I}_1, \mathbf{J}_1, \mathbf{K}_1\}$; the quaternion units $\mathbf{I}_2, \mathbf{J}_2, \mathbf{K}_2$ are from an independent set; e_0, e_1, e_2, e_3 are vector units; $\sigma = -e_1e_2e_3$ is a notational convenience; A, B, C, D are Clifford vectors of the form $A = A_xe_1 + A_ye_2 + A_ze_3$; and a, b, c, d are scalars.

The terms in parentheses are quaternions expressed in scalar plus vector form such as $\mathfrak{A} = a + A\sigma$. In this form a is the scalar part, the part $A\sigma = A_xe_2e_3 + A_ye_3e_1 + A_ze_1e_2$ is an oriented area, and $A = A_xe_1 + A_ye_2 + A_ze_3$ is the vector part, normal to the oriented area.

By setting various of the sixteen constant coefficients to zero, a four-dimensional Clifford value can be set to represent scalars, vectors, oriented areas and volumes, quaternions and complex numbers. The central multiplication of Clifford values can be used to produce central (also called Clifford), outer (wedge, $\wedge$), inner (dot, $\cdot$) and cross ($\times$) products.

Clifford numbers and Clifford's algebra embody in a single coherent form many of the algebraic and geometric entities and operations commonly used.

1.6.1 Notation

Given that a Clifford number of four dimensions can represent a multitude of mathematical and geometric entities, a choice must be made regarding whether to adopt different ways of writing the symbols representing each different entity or not. To write a single Clifford number x as a real number, a complex number, a quaternion, a full Clifford number, a vector, an oriented area, an oriented volume or an oriented entity of the fourth dimension, all distinctly, requires some eight different ways of expressing the same symbol.

Such a system is flawed at the outset. The supposed clarity it is intended to bring is lost immediately any two of these entities are added or multiplied together unless additional special symbols are used for paired or triple combinations. One example is the integrand of the Cauchy integral, being the product between two particular vectors and a bivector (see Sect. 7.3.1). The result is in the form of a sum of complex scalar, oriented area, and oriented fourth-dimensional entity.

Should an extra special symbol be created? Ultimately the answer has to be no, because the number of required symbols grows exponentially. The solution is simply to use for this compound entity the single common symbolic expression for the most general Clifford number with that many number of dimensions. With this approach (really the only practical one) all entities are ultimately rendered using the same typescript, and cannot be distinguished by sight alone.

For much of part II of the book a single typescript is adopted for mathematical and geometric entities of all types. The nature of a particular symbol x must ultimately be determined from the context in which it is found, *locally* only. Nevertheless, some of the loosely standard conventions for particular types of entities are adopted in part I, where it is important to provide a path to make the transition from specific historical conventions to the convention of having no distinct typescripts.

In most of part I, real and complex scalars are rendered in lowercase italic script x, as indicated in Table 1.4. Quaternions are rendered in uppercase Fraktur script $\mathfrak{X}$, except in the case of the four quaternion units which are rendered in uppercase bold script $\mathbf{X}$. In the early chapters where many different entities coexist, general Clifford numbers are rendered in uppercase calligraphic script $\mathcal{X}$.

Ordinary vectors as found in vector calculus are either lowercase bold script $\mathbf{x}$ or uppercase bold script $\mathbf{X}$ to match with either the geometrical convention for a point in space or the electromagnetic convention for a field quantity. The corresponding Clifford vectors x and fields X are not bold. Non-Clifford vector units carry a hat $\hat{\mathbf{x}}$ and Clifford vector units take the form e_x (see Sect. 2.1). No special notation is used for oriented areas, volumes or higher-dimensional entities.

Table 1.4 Typescripts for different mathematical and geometric entities

Entity	Scripts			
Real	$abc \dots xyz$			
Complex	$abc \dots xyz$			
Quaternion	$\mathfrak{ABC} \dots \mathfrak{XYZ}$			
Quaternion unit	$\mathbf{ABC} \dots \mathbf{XYZ}$			
Clifford	$abc \dots xyz$	$\mathcal{ABC} \dots \mathcal{XYZ}$		
Vector	$\mathbf{abc} \dots \mathbf{xyz}$	$\mathbf{ABC} \dots \mathbf{XYZ}$	$abc \dots xyz$	$ABC \dots XYZ$
Vector unit	$\hat{\mathbf{a}}\hat{\mathbf{b}}\hat{\mathbf{c}} \dots \hat{\mathbf{x}}\hat{\mathbf{y}}\hat{\mathbf{z}}$	$e_a e_b e_c \dots e_x e_y e_z$		
Oriented area	$abc \dots xyz$			
Oriented volume	$abc \dots xyz$			
Oriented four-space	$abc \dots xyz$			
Relativistic frames	$\mathbb{ABC} \dots \mathbb{XYZ}$			

A special symbol ∇ is adopted in three dimensions only, to distinguish the vector differential operator:

$$\nabla = \frac{\partial}{\partial x}\hat{\mathbf{x}} + \frac{\partial}{\partial y}\hat{\mathbf{y}} + \frac{\partial}{\partial z}\hat{\mathbf{z}} \tag{1.20}$$

in the context of Gibbs's vector calculus from the Clifford vector differential operator:

$$\boldsymbol{\nabla} = \frac{\partial}{\partial x}e_1 + \frac{\partial}{\partial y}e_2 + \frac{\partial}{\partial z}e_3 \tag{1.21}$$

in the context of Clifford algebra. The distinction is important because the range of multiplication operations which is supported by each is different. The conventional vector operator ∇ supports only scalar, dot and cross products, whereas the Clifford vector operator $\boldsymbol{\nabla}$ supports the full range of vector and Clifford multiplication and all of the variants described in Sect. 2.3.3. See Sect. 5.4 for details.

1.7 Casting Physical Problems into Clifford Algebra

For the purpose of solving physical problems with Clifford's algebra it is necessary to cast whatever quantities are used to describe the physical phenomena involved, and the way those quantities are observed to behave and interact with one another, into a Clifford algebraic formulation.

For example, in the case of the phenomenon of electromagnetism, the quantities of most interest are the sources (the *current density* $\mathbf{J}$ and the *charge density* ρ) and the fields (the *electric field* $\mathbf{E}$ and the *magnetic field* $\mathbf{H}$). In addition it is necessary to include the physical properties of the materials (the *electric permittivity* ϵ and the *magnetic permeability* μ) in those regions where the fields exist.

In order to reproduce Maxwell's original equations [1][3] it is necessary to ensure that the Clifford representation of the electromagnetic field together with the Clifford representation of the partial differential operators (*div* $\nabla\cdot$, *curl* $\nabla\times$ and $\frac{\partial}{\partial t}$) form a correctly matched pair. To adhere to physical behaviour it is necessary to ensure this matched pair produces consistent results when viewed (or mapped) from one moving frame of reference to another. Doing so properly incorporates Einstein's laws of special relativity [21]. See Chap. 4 for details.

Customarily ρ, μ, ϵ and $\frac{\partial}{\partial t}$ are treated as scalars, and $\mathbf{E}, \mathbf{H}, \mathbf{J}$, $\nabla\cdot$ and $\nabla\times$ are treated as vectors. The former are candidates for any of the scalars a, b, c, d in Eq. 1.19, and the latter are candidates for the Clifford vectors A, B, C, D.

When encoded according to the phenomenological considerations the electromagnetic field $\mathcal{F}$ is written in Clifford form as [22]:

$$\mathcal{F} = \sqrt{\mu}\, H\sigma - i\sqrt{\epsilon}\, Ee_0 \tag{1.22}$$

the sources are written:

$$\mathcal{S} = \sqrt{\mu}\, J + \tfrac{i}{\sqrt{\epsilon}}\rho e_0 \tag{1.23}$$

the differential operator is written:

$$\mathcal{D} = \nabla - \tfrac{i}{c} e_0 \tfrac{\partial}{\partial t} \tag{1.24}$$

and Maxwell's equations are written:

$$\mathcal{D}\mathcal{F} = \mathcal{S} \tag{1.25}$$

where $\sigma = -e_1e_2e_3$, $i^2 = -1$, $\nabla = \frac{\partial}{\partial x}e_1 + \frac{\partial}{\partial y}e_2 + \frac{\partial}{\partial z}e_3$ and c is the speed of light. See Sect. 5.1.2 for details.

Once Maxwell's equations are formulated in this manner they can be solved [2, 12] using the Cauchy integral [8][4] (also [9][5]) in its multi-dimensional Cliffordian form [10]. See part II for details.

1.8 Exercises

Q1. Using the algebraic properties of the outer product show that the two evolutions below have the same value in spite of their obviously different geometric shapes.

[3]Section 617.

[4]Equation 6.

[5]Section 6.9

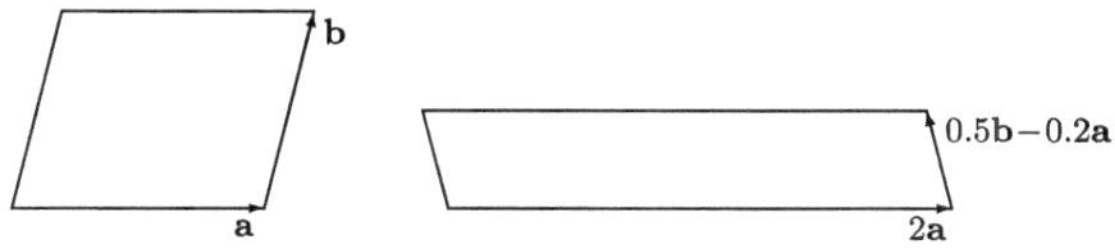

Q2. Show that the two evolutions below are equal by approximating one of them with arbitrarily small parallelograms.

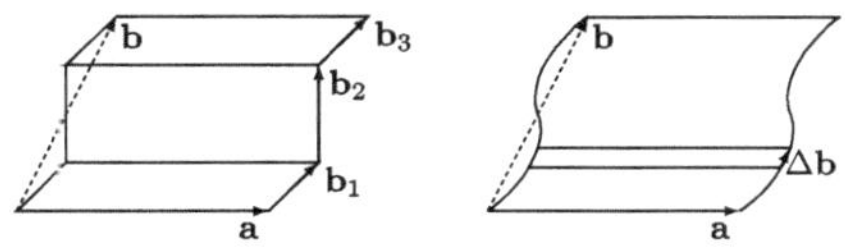

Q3. Calculate the outer, central and inner products for the two vectors $\mathbf{a} = a_x e_1 + a_y e_2 + a_z e_3$ and $\mathbf{b} = b_x e_1 + b_y e_2 + b_z e_3$. Write the individual terms of each product in a 3 × 3 table such as shown below. Compare the three tables and verify that $\mathbf{ab} = \mathbf{a} \wedge \mathbf{b} - \mathbf{a} \cdot \mathbf{b}$.

central ab		b		
		$b_x e_1$	$b_y e_2$	$b_z e_3$
a	$a_x e_1$			
	$a_y e_2$			
	$a_z e_3$			

outer $\mathbf{a} \wedge \mathbf{b}$		b		
		$b_x e_1$	$b_y e_2$	$b_z e_3$
a	$a_x e_1$			
	$a_y e_2$			
	$a_z e_3$			

inner $\mathbf{a} \cdot \mathbf{b}$		b		
		$b_x e_1$	$b_y e_2$	$b_z e_3$
a	$a_x e_1$			
	$a_y e_2$			
	$a_z e_3$			

Q4. Show that under central multiplication the vector units e_1, e_2 and e_3 retain the associative property of the outer product.

Q5. Using Eq. 1.8 show for all quaternion units **I**, **J**, **K** that the constraints of Grassmann's central multiplication apply.

Q6. Show that Hamilton's constraint equations 1.8 for quaternion units follow directly from Clifford's construction of quaternion units in Eq. 1.10.

Q7. Show that $\sigma^2 = 1$, and that σ commutes with vector units e_1, e_2 and e_3 but anticommutes with vector unit e_0.

Q8. Show that Hamilton's constraint equations 1.8 for quaternion units follow directly from the construction of quaternion units in Eq. 1.11.

Q9. Show that the two sets of quaternion units in Eqs. 1.10 and 1.11 are commutative.

Q10. Calculate all sixteen units of Clifford's four-dimensional algebra.

Q11. Calculate the cross product in Cartesian components from Eq. 1.15 and verify from an independent source.

Q12. Substitute $\mathbf{a} = \nabla = \frac{\partial}{\partial x} e_1 + \frac{\partial}{\partial y} e_2 + \frac{\partial}{\partial z} e_3$ into Eq. 1.15 to produce an expression for curl. Expand and verify from an independent source.

Q13. Substitute $\mathbf{a} = \nabla$ into Eq. 1.16 to produce an expression for the Clifford product $\nabla \mathbf{b}$ in terms of div and curl.

Q14. Show that the Clifford product $\mathbf{ab} = \nabla\nabla = \nabla^2$ produces, to within a multiplicative constant, the Laplacian.

Q15. Substitute $\mathcal{D}$ and $\mathcal{F}$ into Eq. 1.25 and use the distributive property to expand into four terms. Then expand the Clifford products ∇H and ∇E as in exercise 13. Separate the result into a system of four equations, so that the terms of each has as a common factor one of the quaternion units from the set $\{\mathbf{1}, \mathbf{I}_2, \mathbf{J}_2, \mathbf{K}_2\}$. Finally verify from an independent source that, to within multiplicative constants, the four equations are the same as Maxwell's equations.

A15. See Eqs. 5.6 and 5.7.

Q16. Show that the Clifford product $\mathcal{D}^2 = \mathcal{D}\mathcal{D}$ produces, to within a multiplicative constant, the d'Alembertian (the scalar wave operator).

Chapter 2
Numbers and Arithmetic

2.1 Structure of Clifford Numbers

Clifford numbers are multi-dimensional entities which follow certain simple rules for the arithmetic operations of addition and multiplication. Each Clifford number is composed of several different *components*. Table 2.1 gives examples of Clifford numbers for dimensions from zero to three.

The number of components in any Clifford number depends on its dimension. In general an m-dimensional Clifford number has 2^m components. Of course, some of these components may be of zero value. For problems in electromagnetism four-dimensional Clifford numbers are required, providing $2^4 = 16$ components. However the electromagnetic field has only six non-zero components, three for the electric part of the field and three for the magnetic part of the field. The other ten components are identically zero.

Each component in a Clifford number is made up from two parts: a *coefficient* and a *unit*.

The coefficient is a numeric entity such as a real or complex number. In algebraic calculations, when the value of the numeric coefficient is unknown, a symbolic variable which represents the value of the coefficient is used instead. It is also possible to use matrices and operators as coefficients. For problems in electromagnetism two types of coefficients are used: complex numbers and partial differential operators. When coefficients combine in operations such as addition and multiplication they behave according to the normal rules for entities of their own particular types.

The unit is a symbolic entity. By convention, the symbol is constructed from the letter e with a subscript attached, either singly or as a product. Products of units follow Clifford's rules (constraint equations):

$$e_r e_s = \begin{cases} -1 & \text{when } r = s \\ -e_s e_r & \text{when } r \neq s \end{cases} \tag{2.1}$$

A. Seagar, *Application of Geometric Algebra to Electromagnetic Scattering*,
DOI 10.1007/978-981-10-0089-8_2

Table 2.1 Examples of Clifford numbers

dimension	grade								no. of components				
	$\bigwedge^0$	$\bigwedge^1$			$\bigwedge^2$			$\bigwedge^3$	$\bigwedge^0$	$\bigwedge^1$	$\bigwedge^2$	$\bigwedge^3$	all
0	$3+4i$								1				1
1	$3+4i$	$+7e_0$							1	1			2
2	$3+4i$	$+7e_0$	$+\pi e_1$		$+ie_0e_1$				1	2	1		4
3	$3+4i$	$+7e_0$	$+\pi e_1$	$+xe_2$	$+ie_0e_1$	$+be_0e_2$	$+ce_1e_2$	$+21.4e_0e_1e_2$	1	3	3	1	8
coefficients	$(3+4i)$	7	π	x	i	b	c	21.4					
units		e_0	e_1	e_2	e_0e_1	e_0e_2	e_1e_2	$e_0e_1e_2$					
components	$(3+4i)$	$7e_0$	πe_1	xe_2	ie_0e_1	be_0e_2	ce_1e_2	$21.4e_0e_1e_2$					

Arbitrarily complicated products of units can always be reduced to the form $\prod_p e_p$ where no two subscripts p are the same.

Note that the product of units is not commutative. From this property it follows that the product of Clifford numbers themselves is not commutative, irrespective of the type of coefficients. Coefficients and units freely commute with one another, and coefficients may or may not commute amongst themselves depending on their nature, e.g. numbers do but matrices don't.

Components are classified into *grades* according to their units. Components with no visible units, such as $(3+4i)$ in Table 2.1, fall into grade zero, denoted symbolically by a large wedge with a superscript zero, i.e. $\bigwedge^0$. Components in grade zero play the geometric role of *scalars*. Components containing a single visible e factor (when reduced according to the rules of Eq. 2.1) e_p, fall into grade one: $\bigwedge^1$. These components play the geometric role of *vectors* (oriented line segments). Components with a pair of visible e factors, $e_r e_s$ where $r \neq s$, fall into grade two, $\bigwedge^2$, and play the role of *bivectors* (oriented areas). Similarly, in $\bigwedge^3$ are components with units $e_q e_r e_s$ which play the role of *trivectors* (oriented volumes). In general components containing ℓ distinct units, such as $\prod_{p=1}^{\ell} e_p$, fall into grade ℓ and play the role of ℓ-vectors.

A linear combination of all ℓ-vectors from all grades is called a *multivector*. Clifford numbers are multivectors.

2.2 Notation

2.2.1 Units

The units in Clifford numbers are written in a variety of forms, in order to simply the notation where possible.

The unit for grade zero $\bigwedge^0$, the grade which accommodates scalars, is variously written as 1, $e_\emptyset$ where $\emptyset$ is the empty set or, as in Table 2.1, is left unwritten. Grassmann called this the *absolute* unit, and classified it as a unit of zeroth order. Note that the absolute unit is strictly not a Clifford unit, and that when written as $e_\emptyset$ does not follow the rules 2.1 for products of Clifford units.

The units of first order, those in grade one $\bigwedge^1$, the grade which accommodates vectors, are written either as the symbol e with integer subscripts, for example e_0, e_1, e_2, or with lower case alphabetic subscripts, for example e_q, e_r, e_s. The alphabetic subscripts are usually taken to represent integers although that is by no means necessary. These units are referred to as *primal* units, being the basis from which all units of higher order are generated. Note that the (first order) primal (vector) unit $e_0 \in \bigwedge^1$ is not the same as the (zeroth order) absolute (scalar) unit $e_\emptyset \in \bigwedge^0$.

Higher order units, for grades beyond $\bigwedge^1$, are either written directly as products of primal units, for example e_0e_1, e_re_s, $e_qe_3e_4$, or written as the symbol e with uppercase alphabetic subscripts, for example e_A, e_B, e_S. The uppercase alphabetic subscripts are taken to represent sets. These sets contain subscripts p for all of the primal units e_p which must be formed into a product to generate the units of higher order. For example:

$$
\begin{array}{lll}
e_A & \text{where } A = \{0, 1, 2\} & \Rightarrow e_A = e_0e_1e_2 \\
e_B & \text{where } B = \{q, r, s\} & \Rightarrow e_B = e_qe_re_s \\
e_S & \text{where } S = \{A, B\} & \Rightarrow e_S = e_Ae_B = e_0e_1e_2e_qe_re_s
\end{array}
\tag{2.2}
$$

These higher order units are called *evolved* units, since the Clifford product which generates them reduces to the outer product (as in Sect. 1.2.2) and thereby to Grassmann's process of evolution. The unit e_S is an evolved unit composed of two factors e_A and e_B. Units e_A and e_B are themselves both evolved units, each composed of three factors. Those six factors are all in grade one (i.e. they are primal units), so they contain no further factors apart from themselves and the absolute unit. Strictly speaking that final (trivial) factorisation means that primal units can also be considered as evolved from the absolute unit as generating element using themselves as the measure of generation.

When units are expanded fully in primal form it is easy to identify the grade to which they belong. When units are written in evolved form, the grade to which the unit belongs is not always immediately apparent.

As sets, the subscripts of all evolved units contain also the empty (null) set $\emptyset$. The evolved unit e_S of Eq. 2.2 could also be expanded as:

$$
e_S \quad \text{where } S = \{A, B\} \Rightarrow e_S = e_\emptyset e_Ae_B = e_\emptyset e_0e_1e_2e_qe_re_s \tag{2.3}
$$

Although it is usually convenient to leave the absolute unit $e_\emptyset = 1$ unwritten, it is important to take the empty set explicitly into account when performing set operations on the subscripts of evolved units, as in Sect. 2.3.3.

2.2.2 Vectors

The notation used here to write Clifford vectors takes the form of a summation over a lowercase (integer) index of components:

$$x = \sum_p x_p e_p \tag{2.4}$$

The x_p are the coefficients, and the e_p are the units, all being primal units (from grade one). The index p varies through some range of integers. For an m-dimensional vector the index plays the role of m integers, one for each of the m units therein. Integer values may well be chosen in the range from 0 to $m-1$ or from 1 to m, according to preference.

For example, a four-dimensional Clifford vector can be written:

$$x = \sum_{p=0}^{3} x_p e_p = x_0 e_0 + x_1 e_1 + x_2 e_2 + x_3 e_3 = \sum_{p=1}^{4} x'_p e'_p = x'_1 e'_1 + x'_2 e'_2 + x'_3 e'_3 + x'_4 e'_4 \tag{2.5}$$

where $x'_p = x_{p-1}$ and $e'_p = e_{p-1}$.

For problems in electromagnetism the unit e_0 is taken to represent the dimension of time or frequency, and the units e_1, e_2, e_3 are taken to represent the three orthogonal dimensions of space.

2.2.3 Numbers

The notation used here to write Clifford numbers more general than vectors takes the form of a summation over an uppercase (set) index of components:

$$x = \sum_S x_S e_S \tag{2.6}$$

The x_S are the coefficients, and the e_S are units from all grades, being either absolute, primal, or evolved. The index S varies through some range of sets. For an m-dimensional number the index plays the role of 2^m sets, one for each of the 2^m units in an m-dimensional Clifford number.

For example, a four-dimensional Clifford number can be written:

$$\begin{aligned} x = \sum_{S \subseteq \{0,1,2,3\}} x_S e_S = x_\emptyset e_\emptyset \\ &+ x_0 e_0 + x_1 e_1 + x_2 e_2 + x_3 e_3 \\ &+ x_{0,1} e_0 e_1 + x_{0,2} e_0 e_2 + x_{0,3} e_0 e_3 + x_{1,2} e_1 e_2 + x_{1,3} e_1 e_3 + x_{2,3} e_2 e_3 \\ &+ x_{0,1,2} e_0 e_1 e_2 + x_{0,1,3} e_0 e_1 e_3 + x_{0,2,3} e_0 e_2 e_3 + x_{1,2,3} e_1 e_2 e_3 \\ &+ x_{0,1,2,3} e_0 e_1 e_2 e_3 \end{aligned} \tag{2.7}$$

where each of the five grades is written on a separate line on the right hand side. The empty (null) set has been included explicitly in the expansion for grade zero, but left unwritten for the other grades.

2.3 Arithmetic Operators

2.3.1 Inversion of Signs

For complex numbers it is possible to invert the sign of either the real or imaginary components separately using the operations of negation and complex conjugation. For Clifford numbers there are many more components and many more possibilities. Table 2.2 shows in addition to the operation of negation, three other operations on Clifford numbers: involution, reversion and Clifford conjugation. All three of these involve inverting the signs of some of the components according to a particular rule or pattern based on the grade of the units in those components, as shown in Table 2.3.

The operation of *involution* (denoted by the symbol '$\neg$') involves changing the sign of every primal unit e_p in every evolved unit e_S. The effect is to multiply every component containing a unit in grade ℓ by the factor of $\xi = (-1)^{\ell}$. The overall effect is to invert the sign of all odd numbered grades, as shown in Table 2.3.

Table 2.2 Operations involving inversion of sign

Operation	$f(x)$	Definition		Effect on primal units		Effect on evolved unit
Identity	x	$x = \sum_S x_S e_S$	$e_S = \prod_{p=1}^{\ell} e_p = \xi e_S$			$\xi = 1$
Negation	$-x$	$-x$ $= \sum_S x_S(-e_S)$	$-e_S = -\prod_{p=1}^{\ell} e_p$ $= \xi e_S$			$\xi = -1$
Involution	$\overset{\neg}{x}$	$\overset{\neg}{x} = \sum_S x_S \overset{\neg}{e_S}$	$\overset{\neg}{e_S} = \prod_{p=1}^{\ell}(-e_p)$ $= \xi e_S$		Change sign	$\xi = (-1)^{\ell}$
Reversion	$\widetilde{x}$	$\widetilde{x} = \sum_S x_S \widetilde{e_S}$	$\widetilde{e_S} = \prod_{p=1}^{\ell} e_{\ell+1-p}$ $= \xi e_S$	Reverse order		$\xi = (-1)^{\ell(\ell-1)/2}$
Clifford conjugation	$\overline{x}$	$\overline{x} = \sum_S x_S \overline{e_S}$	$\overline{e_S} = \prod_{p=1}^{\ell}(-e_{\ell+1-p})$ $= \xi e_S$	Reverse order	Change sign	$\xi = (-1)^{\ell(\ell+1)/2}$

Table 2.3 Periodic behaviour of reversion, involution and Clifford conjugation

grade	0	1	2	3	4	5	6	7	8	9	10	11
involution	+	−	+	−	+	−	+	−	+	−	+	−
reversion	+	+	−	−	+	+	−	−	+	+	−	−
Clifford conjugation	+	−	−	+	+	−	−	+	+	−	−	+

The operation of *reversion* (denoted by the tilde symbol '$\sim$') involves reversing the entire sequence of primal units e_p in every evolved unit e_S. Returning all of the primal units to their original order then requires $\ell(\ell-1)/2$ exchanges of adjacent pairs. The effect is to multiply every component containing a unit in grade ℓ by the factor of $\xi = (-1)^{\ell(\ell-1)/2}$.

The operation of *Clifford conjugation* (denoted by the overbar symbol '$-$') involves both involution and reversion, i.e. both changing the sign of every primal unit e_p and also reversing the order of the entire sequence of primal units. The effect is to multiply every component containing a unit in grade ℓ by the factor of $\xi = (-1)^{\ell(\ell+1)/2}$.

2.3.2 Addition and Subtraction

Addition and subtraction of Clifford numbers is straightforward. The coefficients of components with matching units are simply added or subtracted according to the normal rules for entities of their own particular types. For two Clifford numbers $a = \sum_S a_S e_S$ and $b = \sum_S b_S e_S$, the sum $a + b = c$ is:

$$a + b = \left(\sum_S a_S e_S\right) + \left(\sum_S b_S e_S\right) = \sum_S (a_S + b_S) e_S = \sum_S c_S e_S = c \qquad (2.8)$$

and the difference $a - b = c$ is:

$$a - b = \left(\sum_S a_S e_S\right) - \left(\sum_S b_S e_S\right) = \sum_S (a_S - b_S) e_S = \sum_S c_S e_S = c \qquad (2.9)$$

The index set S is taken to include all of the units in both a and b.

2.3.3 Multiplication

Real, complex and Clifford multiplications have zero, one and two constraint equations respectively. Each time more constraints are applied a new kind of multiplication is created.

For Clifford numbers, in addition to the constraints on the primal units e_r, e_s as in Eq. 2.1, a whole new layer of constraints is applied at the level of the evolved units e_A, e_B. Each different constraint (or set of constraints) leads to a different multiplication.

All of these kinds of multiplication can be defined in terms of a general unconstrained product operation (see [7]), as denoted here by the symbol '$\diamond$'. For two Clifford numbers $a = \sum_A a_A e_A$ and $b = \sum_B b_B e_B$, the general product $a \diamond b = c$ is:

$$a \diamond b = \left(\sum_A a_A e_A\right) \diamond \left(\sum_B b_B e_B\right) = \sum_A \sum_B a_A b_B (e_A \diamond e_B) = \sum_C c_C e_C = c \tag{2.10}$$

The value of the product is determined by the conventional product of its coefficients $c_C = a_A b_B$ according to the normal rules for entities of their own particular types, and the general product of the units $e_C = e_A \diamond e_B$ adhering to the constraint equations for the particular kind of multiplication the general product operation $\diamond$ is intended to deliver.

Table 2.4 shows seven different kinds of product, all distinguished by the way they adhere to the general constraint equation:

$$e_A \diamond e_B = \begin{cases} e_A e_B & \text{when } \checkmark \\ 0 & \text{otherwise} \end{cases} \tag{2.11}$$

The tick $\checkmark$ is determined for each of the kinds of multiplication according to such relationships as whether the unit e_A contains all or none of the primal units e_p found in unit e_B, and/or *vice versa* or not. The empty set $\emptyset$ corresponding to the absolute unit $e_\emptyset = 1$ is taken into account as part of the determination.

Table 2.4 shows the constraint equations for the different kinds of multiplication in set notation, and Table 2.5 shows the result of applying the constraint equations to a product of two-dimensional Clifford numbers.

Table 2.4 Constraint equations for seven different kinds of Clifford multiplication

Product	$a \diamond b$	Constraint equations				
		Excludes $B \cap A = \emptyset$	Either includes $B \overset{\supseteq}{\subseteq} A$		Vector $\lvert B\rvert\lvert A\rvert = 1$	Non-vector $\lvert B\rvert\lvert A\rvert \neq 1$
			Includes $B \supseteq A$	Includes $B \subseteq A$		
Central	$a\, b$	✓	✓	✓	✓	✓
Outer	$a \wedge b$	✓			✓	✓
Inner	$\overline{a} \vee b$		✓	✓	✓	✓
Left inner	$\overline{a} \lrcorner b$		✓		✓	✓
Right inner	$a \llcorner \overline{b}$			✓	✓	✓
Scalar	$(\overline{a}, b)$		✓ + ✓		✓	✓
Dot	$\overline{a} \cdot b$		✓ + ✓ + ✓			

Table 2.5 Multiplication of two-dimensional Clifford numbers a and b

		Set	Units															
		A	1	1	1	1	e_0	e_0	e_0	e_0	e_1	e_1	e_1	e_1	e_0e_1	e_0e_1	e_0e_1	e_0e_1
		B	1	e_0	e_1	e_0e_1	1	e_0	e_1	e_0e_1	1	e_0	e_1	e_0e_1	1	e_0	e_1	e_0e_1
Product	$a \diamond b$	C	1	e_0	e_1	e_0e_1	e_0	-1	e_0e_1	$-e_1$	e_1	$-e_0e_1$	-1	e_0	e_0e_1	e_1	$-e_0$	-1
Central (Clifford)	$a\,b$	$\sum_A \sum_B$	√	√	√	√	√	√	√	√	√	√	√	√	√	√	√	√
Outer (wedge)	$a \wedge b$	$\sum_A \sum_{B \cap A = \emptyset}$	√	√	√	√	√		√		√	√			√			
Inner	$\overline{a} \vee b$	$\sum_A \sum_{B \underset{\subseteq}{\supseteq} A}$	√	√	√	√	√	√		√	√		√	√	√	√	√	√
Left inner	$\overline{a} \lrcorner b$	$\sum_A \sum_{B \supseteq A}$	√	√	√	√		√		√			√	√				√
Right inner	$a \llcorner \overline{b}$	$\sum_A \sum_{B \subseteq A}$	√				√	√			√		√		√	√	√	√
Scalar	$(\overline{a}, b)$	$\sum_A \sum_{B=A}$	√					√					√					√
Dot	$\overline{a} \cdot b$	$\sum_{A \in \wedge^1} \sum_{B=A}$						√					√					

The *central* product ab (Clifford multiplication, denoted by no visible symbol) is the most general, with no constraint equations at the level of the evolved units. All of the other products derived from the general product are contained within the central product.

The *outer* product $a \wedge b$ (wedge product, denoted by the symbol '$\wedge$') is constrained to produce non-zero values only when both factors e_A, e_B are composed of units, none of which match any primal unit in the other.

The *inner* product $\overline{a} \vee b$ (denoted by the symbol '$\vee$') is constrained to produce non-zero values only when either of the factors e_A, e_B is composed of units, all of which match any unit in the other.

Grassmann called the outer and inner products by those names because they give non-zero results only if the two factors e_A, e_B either lie completely *outside* the linear span of the primal units contained in each other, or one lies completely *inside* the linear span of the primal units contained in the other.

The *left inner* product $\overline{a} \lrcorner b$ (denoted by the symbol '$\lrcorner$') is constrained to produce non-zero values only when the leftmost of the factors e_A, e_B is composed of units, all of which match any unit in the other.

The *right inner* product $a \llcorner \overline{b}$ (denoted by the symbol '$\llcorner$') is constrained to produce non-zero values only when the rightmost of the factors e_A, e_B is composed of units, all of which match any unit in the other.

The *scalar* product $(\overline{a}, b)$ (denoted by the parenthesis-comma-parenthesis notation '(,)') is constrained to produce non-zero values only when both of the factors e_A, e_B are composed of units, all of which match any unit in the other.

For the three inner products, and for the scalar product, the conjugation can be moved by substituting $a' = \overline{a}$ or $b' = \overline{b}$, which then leads to a product of factors $e_A e_B = \overline{e_{A'}} e_B$ or $e_A e_B = e_A \overline{e_{B'}}$. For the scalar product the factor to which the conjugation is applied is of little consequence since the result, always a scalar, is the same in either case.

The *dot* product $\overline{a} \cdot b$ (denoted by the symbol '$\cdot$') is constrained to produce non-zero values only when both of the factors e_A, e_B are composed of the one and same primal unit. For vectors **a**, **b** the value is given in terms of Clifford multiplication by Eq. 1.14, and for primal units e_r, e_s the value is given in Table 1.2. The dot product is a special case of the inner and scalar products, just for vectors.

The seven products listed in Table 2.5 are not all independent. The left and right inner products are defined in terms of the outer and scalar products by the two duality relationships:

$$\begin{cases} (u \lrcorner x, y) = (x, u \wedge y) \\ (x \llcorner u, y) = (x, y \wedge u) \end{cases} \tag{2.12}$$

These relationships hold for both the Hamilton, Maxwell, Grassmann, Clifford convention $e_r^2 = -1$, and for the alternative convention $e_r^2 = +1$. For the alternative convention the products in Table 2.5 employing Clifford conjugation must be modified so as to use reversion instead. It is important when reading published literature relating to Clifford algebra to determine early which convention has been used so as to avoid confusion.

For problems in electromagnetism as treated here it is the unconstrained Clifford or central multiplication which is of primary importance. For problems in differential geometry it is the outer and left inner products which play the major roles.

2.3.4 Application

Figure 2.1 shows two Clifford vectors:

$$\begin{cases} a = a_{\|} + a_{\perp} = e_1 \\ b = b_{\|} + b_{\perp} = \frac{1}{2}e_1 + \frac{\sqrt{3}}{2}e_2 \end{cases} \tag{2.13}$$

which are each broken into two components, parallel ($\|$) and perpendicular ($\perp$) to the other:

$$\begin{cases} a_{\|} = \frac{1}{4}e_1 + \frac{\sqrt{3}}{4}e_2 \\ a_{\perp} = \frac{3}{4}e_1 - \frac{\sqrt{3}}{4}e_2 \end{cases} \quad \text{and} \quad \begin{cases} b_{\|} = \frac{1}{2}e_1 \\ b_{\perp} = \frac{\sqrt{3}}{2}e_2 \end{cases} \tag{2.14}$$

The vector $a_{\|}$ is the component of a parallel to the direction of b, and the vector $a_{\perp}$ is the component of a perpendicular to the direction of b. The vectors $b_{\|}$ and $b_{\perp}$ behave in a similar way with respect to the direction of a.

By direct calculation the outer product:

$$a_{\perp} \wedge b = a \wedge b = a \wedge b_{\perp} \tag{2.15}$$

depends only on the perpendicular components of the two vectors. By contrast, the inner product:

$$a_{\|} \vee b = a \vee b = a \vee b_{\|} \tag{2.16}$$

depends only on the parallel components of the two vectors.

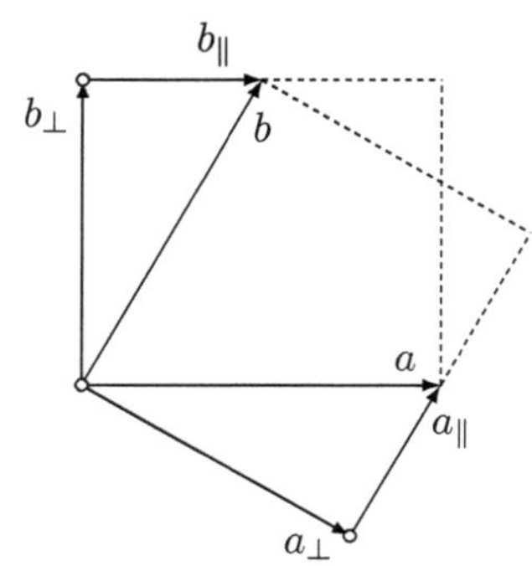

Fig. 2.1 Vectors a and b broken into components perpendicular and parallel to one another

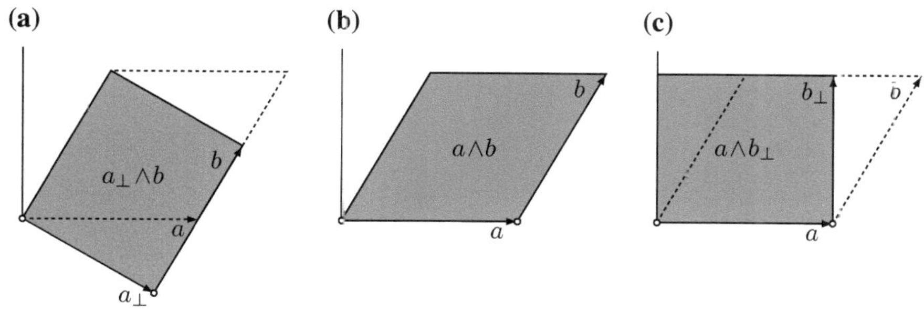

Fig. 2.2 Torque (*shaded*) calculated from outer product of (**a**) perpendicular component $a_\perp$ of force $a = F$ with whole lever-arm $b = d$, of (**b**) whole force $a = F$ with whole lever-arm $b = d$, and of (**c**) whole force $a = F$ with perpendicular component $b_\perp$ of lever-arm $b = d$

In the case that the vector a represents a force F applied to the tip of a lever-arm, and the vector b represents the displacement d from the tip of the level-arm to its pivot, then the geometrical interpretation shown in Fig. 2.2b for the outer product $a \wedge b = F \wedge d$, is the resulting torque as given in value by the area of the shaded portion of the figure. Figure 2.2a shows the case taking the perpendicular component of force F with the whole of the displacement d, and Fig. 2.2c shows the case taking the whole force F with the perpendicular component of the displacement d. In all cases the size of the shaded portion of the figures (i.e. the amount of torque) is the same.

In the case that the vector a represents a force F applied to an object, and the vector b represents the displacement d of that object under the influence of the force, then the geometrical interpretation shown in Figs. 2.3b and 2.4b for the inner product

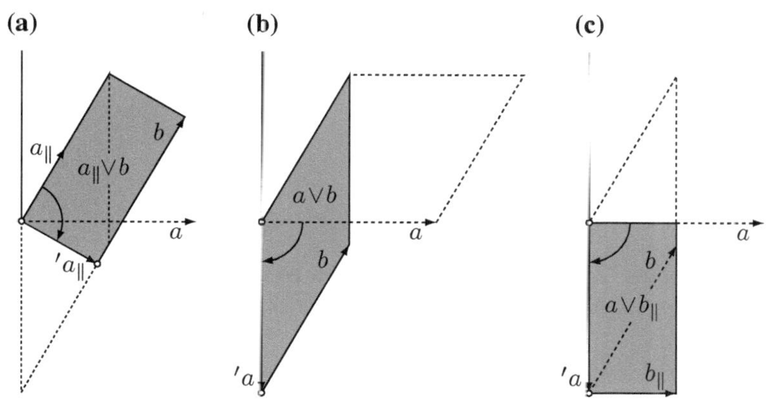

Fig. 2.3 Work (*shaded*) calculated from inner product of (**a**) parallel component $a_\parallel$ of force $a = F$ with whole displacement $b = d$, of (**b**) whole force $a = F$ with whole displacement $b = d$, and of (**c**) whole force $a = F$ with parallel component $b_\parallel$ of displacement $b = d$. The inner product with vector a is converted to an outer product with rotated vector $'a$

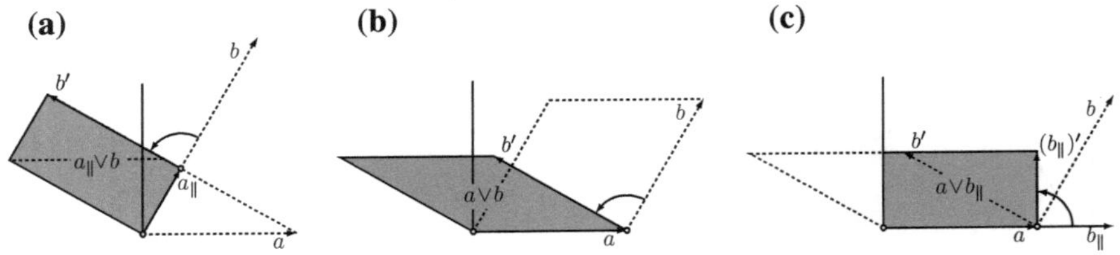

Fig. 2.4 Work (*shaded*) calculated from inner product of (**a**) parallel component $a_{\|}$ of force $a = F$ with whole displacement $b = d$, of (**b**) whole force $a = F$ with whole displacement $b = d$, and of (**c**) whole force $a = F$ with parallel component $b_{\|}$ of displacement $b = d$. The inner product with vector b is converted to an outer product with rotated vector b'

$a \vee b = F \vee d$, is the resulting work as given in value by the area of the shaded portion of the figures. Figures 2.3a and 2.4a show the case taking the parallel component of force F with the whole of the displacement d, and Figs. 2.3c and 2.4c shows the case taking the whole force F with the parallel component of the displacement d. In all cases the size of the shaded portion of the figures (i.e. the amount of work) is the same.

In Fig. 2.3b the vector $'a$ is constructed by rotating the vector a by $-90°$ when angles are measured in a positive sense through the smallest arc from a to b. This converts the inner product of a and b into an evolution (outer product) of $'a$ and b. Similar conversions have been made in Fig. 2.3a, c.

In Fig. 2.4b the vector b' is constructed by rotating the vector b by $+90°$. This converts the inner product of a and b into an evolution (outer product) of a and b'. Similar conversions have been made in Fig. 2.4a, c.

The torque is a bivector and the work is a scalar. Calculating the central product ab produces the torque in grade two of the result and produces the work in grade zero of the result. Depending on the physical situation it may be one or other of these components which is of most interest.

2.4 Numerical Implementation

2.4.1 Dimensional Recursion

A Clifford number $\mathcal{Z}$ of dimension $m > 0$ can be constructed by combining two Clifford numbers $\mathcal{X}$ and $\mathcal{Y}$ of dimension $m - 1$ with the unit e_m for dimension m:

$$\mathcal{Z}_{(m)} = \mathcal{X}_{(m-1)} + \mathcal{Y}_{(m-1)} e_m \tag{2.17}$$

Table 2.6 Recursive construction of Clifford numbers

Dimension	Number		Involution of number	
m	$\mathcal{Z}_{(m)}$	$\mathcal{X}_{(m-1)} + \mathcal{Y}_{(m-1)} e_m$	$\overline{\mathcal{Z}}_{(m)}$	$\overline{\mathcal{X}}_{(m-1)} - \overline{\mathcal{Y}}_{(m-1)} e_m$
$m-1$	$\mathcal{Z}_{(m-1)}$	$\mathcal{X}_{(m-2)} + \mathcal{Y}_{(m-2)} e_{m-1}$	$\overline{\mathcal{Z}}_{(m-1)}$	$\overline{\mathcal{X}}_{(m-2)} - \overline{\mathcal{Y}}_{(m-2)} e_{m-1}$
...	...	...	...	...
2	$\mathcal{Z}_{(2)}$	$\mathcal{X}_{(1)} + \mathcal{Y}_{(1)} e_2$	$\overline{\mathcal{Z}}_{(2)}$	$\overline{\mathcal{X}}_{(1)} - \overline{\mathcal{Y}}_{(1)} e_2$
1	$\mathcal{Z}_{(1)}$	$\mathcal{X}_{(0)} + \mathcal{Y}_{(0)} e_1$	$\overline{\mathcal{Z}}_{(1)}$	$\overline{\mathcal{X}}_{(0)} - \overline{\mathcal{Y}}_{(0)} e_1$
0	$\mathcal{Z}_{(0)}$	$x + iy$	$\overline{\mathcal{Z}}_{(0)}$	$x + iy$

The process can be repeated for $\mathcal{X}$ and $\mathcal{Y}$ with unit e_{m-1}, and for all lower dimensions down to Clifford numbers of dimension zero, as shown in Table 2.6. At that level, a Clifford number is simply constructed from a single complex number $z = x + iy$:

$$\mathcal{Z}_{(0)} = x + iy \tag{2.18}$$

Note that the involution operation (in the two rightmost columns) plays a role for the primal units e_p of Clifford numbers similar to the role the complex conjugate operation plays for the imaginary unit i of complex numbers. However, the involution does not operate on the imaginary unit i, so the involution of a zero-dimensional Clifford number is the same as the original number. (This is similar to complex numbers, where the complex conjugate of a real number is the same as the original number).

2.4.2 Recursive Product

The product of Clifford numbers of any dimension can be recursively defined in terms of Clifford numbers of lower dimension, in a form similar to the (more familiar) product of complex numbers. For complex numbers $u = v + iw$ and $z = x + iy$ the product is:

$$uz = (v + iw)(x + iy) = [vx - wy] + i[vy + wx] \tag{2.19}$$

For m-dimensional Clifford numbers $\mathcal{U}_{(m)} = \mathcal{V}_{(m-1)} + \mathcal{W}_{(m-1)} e_m$ and $\mathcal{Z}_{(m)} = \mathcal{X}_{(m-1)} + \mathcal{Y}_{(m-1)} e_m$ the product is:

$$\begin{aligned}\mathcal{U}_{(m)}\mathcal{Z}_{(m)} &= (\mathcal{V}_{(m-1)} + \mathcal{W}_{(m-1)}e_m)(\mathcal{X}_{(m-1)} + \mathcal{Y}_{(m-1)}e_m) \\ &= [\mathcal{V}_{(m-1)}\mathcal{X}_{(m-1)} + \mathcal{W}_{(m-1)}e_m\mathcal{Y}_{(m-1)}e_m] \\ &\quad + [\mathcal{V}_{(m-1)}\mathcal{Y}_{(m-1)}e_m + \mathcal{W}_{(m-1)}e_m\mathcal{X}_{(m-1)}]\end{aligned} \tag{2.20}$$

Changing the order of $e_m\mathcal{X}_{(m-1)}$ to $\mathcal{X}_{(m-1)}e_m$ introduces minus signs in all odd grades of $\mathcal{X}$. The same effect can be achieved by applying the involution operator (as described in Sect. 2.3.1) so that:

$$e_m \overline{\mathcal{X}}_{(m-1)} = \mathcal{X}_{(m-1)}e_m \tag{2.21}$$

Multiplying both sides from left and right by e_m now gives:

$$\begin{aligned} e_m e_m \overline{\mathcal{X}}_{(m-1)} e_m &= e_m\mathcal{X}_{(m-1)}e_m e_m \\ \overline{\mathcal{X}}_{(m-1)} e_m &= e_m\mathcal{X}_{(m-1)}\end{aligned} \tag{2.22}$$

Using this result for both $e_m\mathcal{X}_{(m-1)}$ and $e_m\mathcal{Y}_{(m-1)}$ in Eq. 2.20 gives:

$$\begin{aligned}\mathcal{U}_{(m)}\mathcal{Z}_{(m)} &= [\mathcal{V}_{(m-1)}\mathcal{X}_{(m-1)} + \mathcal{W}_{(m-1)} \overline{\mathcal{Y}}_{(m-1)} e_m e_m] \\ &\quad + [\mathcal{V}_{(m-1)}\mathcal{Y}_{(m-1)}e_m + \mathcal{W}_{(m-1)} \overline{\mathcal{X}}_{(m-1)} e_m] \\ &= [\mathcal{V}_{(m-1)}\mathcal{X}_{(m-1)} - \mathcal{W}_{(m-1)} \overline{\mathcal{Y}}_{(m-1)}] \\ &\quad + [\mathcal{V}_{(m-1)}\mathcal{Y}_{(m-1)} + \mathcal{W}_{(m-1)} \overline{\mathcal{X}}_{(m-1)}]e_m\end{aligned} \tag{2.23}$$

Comparison of Eq. 2.23 for the multiplication of Clifford numbers to the case for multiplication of complex numbers (Eq. 2.19) shows strong similarities. Structurally the equations are the same, with the only added complication in the case of Clifford multiplication being the need to apply the involution operation to some of the terms. The simple form of Eq. 2.23 can be exploited to develop simple (although recursive) computer implementations of Clifford multiplication, as shown in Sect. 2.4.3.

2.4.3 Machine Computation

Currently no computer languages exist which support generic data types for Clifford numbers. In order to perform machine-based arithmetic operations with Clifford numbers it is necessary both to design a suitable data construct which serves to represent them, and to develop accompanying computer routines which serve to operate on that construct.

The recursive representation of Clifford numbers described in Sect. 2.4.1 exhibits the form of a binary tree in which a number $\mathcal{Z}$ of dimension $m > 0$ consists of an ordered pair of elements (numbers) $\mathcal{X}, \mathcal{Y}$ each of dimension one lower, and each branch of the tree terminates at the lowest dimension $m = 0$ with a single element (number) $\mathcal{Z} = a_0$, where a_0 is a complex number:

$$\mathcal{Z} := (\mathcal{X}\ \mathcal{Y}) \qquad m > 0 \tag{2.24}$$

$$\mathcal{Z} := a_0 \qquad m = 0 \tag{2.25}$$

Substituting two elements in the form of Eq. 2.25 into Eq. 2.24 gives the representation of a Clifford number of dimension $m = 1$:

$$\mathcal{Z} := (\underset{\mathcal{X}}{a_0}\ \underset{\mathcal{Y}}{a_1}) \qquad m = 1 \tag{2.26}$$

where now both a_0 and a_1 are complex numbers. Similarly, substituting two elements in the form of Eq. 2.26 into Eq. 2.24 gives the representation of a Clifford number of dimension $m = 2$. Repeating the substitution recursively gives the representations of Clifford numbers for successively higher dimensions. The representation in tree structure of Clifford numbers $\mathcal{Z}$ of dimension $m \leq 4$ is shown in Table 2.7.

The tree structure contains only the coefficients a_S of the components of a Clifford number. The units e_S need never be recorded explicitly. They can always be determined from the position of the coefficients in the tree.

Multiplication as described in Sect. 2.4.2, of Clifford numbers represented in the tree structure, is only a little more complicated than multiplying two complex num-

Table 2.7 Machine representation in tree structure of Clifford numbers of dimension m from zero to four

$\mathcal{Z} := a_0$	$m = 0$
$\mathcal{Z} := (\underset{\mathcal{X}}{a_0}\ \underset{\mathcal{Y}}{a_1})$	$m = 1$
$\mathcal{Z} := (\underbrace{(a_0\ a_1)}_{\mathcal{X}}\ \underbrace{(a_2\ a_3)}_{\mathcal{Y}})$	$m = 2$
$\mathcal{Z} := (\underbrace{((a_0\ a_1)(a_2\ a_3))}_{\mathcal{X}}\ \underbrace{((a_4\ a_5)(a_6\ a_7))}_{\mathcal{Y}})$	$m = 3$
$\mathcal{Z} := (\underbrace{(((a_0\ a_1)(a_2\ a_3))((a_4\ a_5)(a_6\ a_7)))}_{\mathcal{X}}\ \underbrace{(((a_8\ a_9)(a_{10}\ a_{11}))((a_{12}\ a_{13})(a_{14}\ a_{15})))}_{\mathcal{Y}})$	$m = 4$

Table 2.8 Recursive multiplication of Clifford numbers in REDUCE code

```
procedure cmulx(U,Z);
if tlist(U)=1 and tlist(Z)=1
then list(subx(cmulx(first(U),first(Z)),cmulx(second(U),invc(second(Z)))),
          addx(cmulx(first(U),second(Z)),cmulx(second(U),invc(first(Z)))))
else if tlist(U)=0 and tlist(Z)=0
     then U*Z else write "bad arguments to cmulx";
```

bers. By way of example Table 2.8 shows a computer implementation "cmulx" of Eq. 2.23 in the REDUCE programming language, for multiplying two Clifford numbers "U" and "Z" of equal but arbitrary dimension. This implementation invokes a recursive call on itself, along with calls to companion routines "subx", "addx" and "invc", as described in appendix A, for subtraction, addition and involution. The actual multiplication is performed by the single REDUCE expression "U*Z".

Multiplying Clifford numbers of different dimensions is only slightly more complicated. Routines to do so are listed in appendix A, along with brief descriptions of the purpose of each routine.

Examples of machine computations on Clifford numbers with predominantly integer coefficients (for simplicity of reading) are shown in Table 2.9. Note that examples 2 and 5, of multiplication on one-dimensional Clifford numbers, behave according to the same rules as complex multiplication. Note also that the routines

Table 2.9 Examples of machine computations on Clifford numbers

```
 1: printcliff(cmulx(2,5));
    10
 2: printcliff(cmulx({1,2},{3,4}));
    10*e(0)-5
 3: u:={v,w};
    u:={v,w}
 4: z:={x,y};
    z:={x,y}
 5: printcliff(cmulx(u,z));
    e(0)*(v*y+w*x)+v*x-w*y
 6: printcliff(cmulx({{1,2},{3,4}},{{5,6},{7,8}}));
    30*e(1)+24*e(0)*e(1)+12*e(0)-60
 7: printcliff(cmulx({{{1,2},{3,4}},{{5,6},{7,8}}},
                     {{{9,10},{11,12}},{{13,14},{15,16}}}));
    -38*e(2)-66*e(1)*e(2)+246*e(1)+204*e(0)*e(2)+192*e(0)*e(1)*e(2)
    -152*e(0)*e(1)-220*e(0)-218
 8: z:=cmulx({{{1,2},{3,4}},{{5,6},{7,8}}},
             {{{9,10},{11,12}},{{13,14},{15,16}}});
    z:={{{-218,-220},{246,-152}},{{-38,204},{-66,192}}}
 9: printcliff(z);
    -38*e(2)-66*e(1)*e(2)+246*e(1)+204*e(0)*e(2)+192*e(0)*e(1)*e(2)
    -152*e(0)*e(1)-220*e(0)-218
10: dimc(z);
    3
11: printcliff(invc(z));
    38*e(2)-66*e(1)*e(2)-246*e(1)+204*e(0)*e(2)-192*e(0)*e(1)*e(2)
    -152*e(0)*e(1)+220*e(0)-218
12: printcliff(gradec(z,0));
    -218
13: printcliff(gradec(z,1));
    -38*e(2)+246*e(1)-220*e(0)
14: printcliff(gradec(z,2));
    -66*e(1)*e(2)+204*e(0)*e(2)-152*e(0)*e(1)
15: printcliff(gradec(z,3));
    192*e(0)*e(1)*e(2)
```

Table 2.10 Symbolic verification of Clifford product duality relationship from first line of Eq. 2.12

```
1: m:=4;
   m:=4
2: x:=makecliff(m,x);
   x:={{{{x0,x1},{x2,x3}},{{x4,x5},{x6,x7}}},
       {{{x8,x9},{x10,x11}},{{x12,x13},{x14,x15}}}}
3: y:=makecliff(m,y);
   y:={{{{y0,y1},{y2,y3}},{{y4,y5},{y6,y7}}},
       {{{y8,y9},{y10,y11}},{{y12,y13},{y14,y15}}}}
4: u:=makecliff(m,u);
   u:={{{{u0,u1},{u2,u3}},{{u4,u5},{u6,u7}}},
       {{{u8,u9},{u10,u11}},{{u12,u13},{u14,u15}}}}
5: lhs:=printcliff(skalar(left(u,x),y));
   lhs:=u0*x0*y0+u0*x1*y1+u0*x10*y10+u0*x11*y11+u0*x12*y12+u0*x13*y13+
   u0*x14*y14+u0*x15*y15+u0*x2*y2+u0*x3*y3+u0*x4*y4+u0*x5*y5+u0*x6*y6
   +u0*x7*y7+u0*x8*y8+u0*x9*y9+u1*x1*y0+u1*x11*y10+u1*x13*y12+u1*x15*
   y14+u1*x3*y2+u1*x5*y4+u1*x7*y6+u1*x9*y8+u10*x10*y0+u10*x11*y1-u10*
   x14*y4-u10*x15*y5+u11*x11*y0-u11*x15*y4+u12*x12*y0+u12*x13*y1+u12*
   x14*y2+u12*x15*y3+u13*x13*y0+u13*x15*y2+u14*x14*y0-u14*x15*y1+u15*
   x15*y0+u2*x10*y8-u2*x11*y9+u2*x14*y12-u2*x15*y13+u2*x2*y0-u2*x3*y1+
   u2*x6*y4-u2*x7*y5+u3*x11*y8+u3*x15*y12+u3*x3*y0+u3*x7*y4+u4*x12*y8
   -u4*x13*y9-u4*x14*y10+u4*x15*y11+u4*x4*y0-u4*x5*y1-u4*x6*y2+u4*x7*
   y3+u5*x13*y8-u5*x15*y10+u5*x5*y0-u5*x7*y2+u6*x14*y8+u6*x15*y9+u6*
   x6*y0+u6*x7*y1+u7*x15*y8+u7*x7*y0-u8*x10*y2+u8*x11*y3-u8*x12*y4+u8
   *x13*y5+u8*x14*y6-u8*x15*y7+u8*x8*y0-u8*x9*y1-u9*x11*y2-u9*x13*y4+
   u9*x15*y6+u9*x9*y0
6: rhs:=printcliff(skalar(x,outer(u,y)));
   rhs:=u0*x0*y0+u0*x1*y1+u0*x10*y10+u0*x11*y11+u0*x12*y12+u0*x13*y13+
   u0*x14*y14+u0*x15*y15+u0*x2*y2+u0*x3*y3+u0*x4*y4+u0*x5*y5+u0*x6*y6
   +u0*x7*y7+u0*x8*y8+u0*x9*y9+u1*x1*y0+u1*x11*y10+u1*x13*y12+u1*x15*
   y14+u1*x3*y2+u1*x5*y4+u1*x7*y6+u1*x9*y8+u10*x10*y0+u10*x11*y1-u10*
   x14*y4-u10*x15*y5+u11*x11*y0-u11*x15*y4+u12*x12*y0+u12*x13*y1+u12*
   x14*y2+u12*x15*y3+u13*x13*y0+u13*x15*y2+u14*x14*y0-u14*x15*y1+u15*
   x15*y0+u2*x10*y8-u2*x11*y9+u2*x14*y12-u2*x15*y13+u2*x2*y0-u2*x3*y1+
   u2*x6*y4-u2*x7*y5+u3*x11*y8+u3*x15*y12+u3*x3*y0+u3*x7*y4+u4*x12*y8
   -u4*x13*y9-u4*x14*y10+u4*x15*y11+u4*x4*y0-u4*x5*y1-u4*x6*y2+u4*x7*
   y3+u5*x13*y8-u5*x15*y10+u5*x5*y0-u5*x7*y2+u6*x14*y8+u6*x15*y9+u6*
   x6*y0+u6*x7*y1+u7*x15*y8+u7*x7*y0-u8*x10*y2+u8*x11*y3-u8*x12*y4+u8
   *x13*y5+u8*x14*y6-u8*x15*y7+u8*x8*y0-u8*x9*y1-u9*x11*y2-u9*x13*y4+
   u9*x15*y6+u9*x9*y0
7: lhs-rhs;
   0
```

operate equally well with numeric entities (as in example 2) and symbolic entities (as in example 5).

An example of machine computations to verify the first of the duality relationships from Eq. 2.12 is shown in Table 2.10. This example serves as quite a good check on the implementation of some of the different kinds of products.

2.5 Exercises

Q1. Show that the operations of involution, reversion and Clifford conjugation are all their own inverses.

Q2. Show that the application of any two of the operations of involution, reversion and Clifford conjugation has the same effect as the one not used.

Q3. Calculate the involution, reversion and Clifford conjugation of the number $e_1e_3e_3$ by applying the effects described in Table 2.2 on the primal units directly and then rearranging them into their original order.

Q4. Show that reversing the order of the entire sequence of primal units e_p in every evolved unit e_S in grade ℓ has the effect of multiplying by the factor of $\xi = (-1)^{\ell(\ell-1)/2}$.

Q5. Calculate the sum $a + b$ of the two Clifford numbers $a = 2 - 1e_1$ and $b = 1 + 3e_1 + 2e_2 + 4e_1e_2$.

A5. $3 + 2e_1 + 2e_2 + 4e_1e_2$.

Q6. Calculate the difference $a - b$ of the two Clifford numbers $a = 2 - 1e_1$ and $b = 1 + 3e_1 + 2e_2 + 4e_1e_2$.

A6. $1 - 4e_1 - 2e_2 - 4e_1e_2$.

Q7. Calculate the central product ab of the two Clifford numbers $a = 2 - 1e_1$ and $b = 1 + 3e_1 + 2e_2 + 4e_1e_2$.

A7. $5 + 5e_1 + 8e_2 + 6e_1e_2$.

Q8. Calculate the outer product $a \wedge b$ of the two Clifford numbers $a = 2 - 1e_1$ and $b = 1 + 3e_1 + 2e_2 + 4e_1e_2$.

A8. $2 + 5e_1 + 4e_2 + 6e_1e_2$.

Q9. Calculate the inner product $a \vee b$ of the two Clifford numbers $a = 2 - 1e_1$ and $b = 1 + 3e_1 + 2e_2 + 4e_1e_2$.

A9. $-1 + 7e_1 + 8e_1e_2$.

Q10. Calculate the left inner product $a \rfloor b$ of the two Clifford numbers $a = 2 - 1e_1$ and $b = 1 + 3e_1 + 2e_2 + 4e_1e_2$.

A10. $-1 + 6e_1 + 8e_1e_2$.

Q11. Calculate the right inner product $a \llcorner b$ of the two Clifford numbers $a = 2 - 1e_1$ and $b = 1 + 3e_1 + 2e_2 + 4e_1e_2$.

A11. $-1 - 1e_1$.

Q12. Calculate the scalar product (a, b) of the two Clifford numbers $a = 2 - 1e_1$ and $b = 1 + 3e_1 + 2e_2 + 4e_1e_2$.

A12. -1.

Q13. Calculate the dot product $a \cdot b$ of the two Clifford numbers $a = 2 - 1e_1$ and $b = 1 + 3e_1 + 2e_2 + 4e_1e_2$.

A13. -3.

Q14. Implement using a computer language of your choice, software to multiply two arbitrary Clifford numbers using the central product.

A14. Check your software against question 7.

Q15. Implement using a computer language of your choice, software to multiply two arbitrary Clifford numbers using the outer product.
A15. Check your software against question 8.
Q16. Implement using a computer language of your choice, software to multiply two arbitrary Clifford numbers using the inner product.
A16. Check your software against question 9.
Q17. Implement using a computer language of your choice, software to multiply two arbitrary Clifford numbers using the left inner product.
A17. Check your software against question 10.
Q18. Implement using a computer language of your choice, software to multiply two arbitrary Clifford numbers using the right inner product.
A18. Check your software against question 11.
Q19. Implement using a computer language of your choice, software to multiply two arbitrary Clifford numbers using the scalar product.
A19. Check your software against question 12.
Q20. Implement using a computer language of your choice, software to multiply two arbitrary Clifford numbers using the dot product.
A20. Check your software against question 13.
Q21. Using your implementations from questions 16 to 19 to verify numerically or symbolically the first duality relationship from Eq. 2.12 for dimension $m = 4$ as in Table 2.10.
Q22. Using your implementations from questions 16 to 19 to verify numerically or symbolically the second duality relationship from Eq. 2.12 for dimensions m from 3 to 5.

Chapter 3
Geometry

Clifford called the algebra he developed, based on Grassmann's central product and two sets of Hamilton's quaternions, a *geometric algebra* [3]. Using Clifford's algebra geometric operations such as translation, reflection, projection and rotation can be carried algebraically on geometric entities such as vectors and bivectors represented in the form of Clifford numbers.

For the purposes of particular applications it is usual to embed a given problem into whatever coordinate system seems most appropriate. In some cases it is convenient to subdivide the surface of an object into a set of matching elements (tiles), and to perform some of the calculations using a coordinate system local to each element. The local results may subsequently be transformed into a common coordinate system so that they can be used together to form a solution. This requires simple rigid body transformations, i.e. translation and rotation.

In the Clifford framework translation of vectors and fields can be effected by adding a constant vector to the independent variable, and rotation can be achieved by employing a rotation operator constructed specifically for that purpose. Rotation operators can be constructed from a pair of reflection operators.

In many problems it is useful to split vectors and fields into a mixture of different complementary components (e.g. normal and tangential) by employing two or more projection operators. In some cases it is appropriate to construct the projection operators from a single reflection operator.

3.1 Reflection, Projection and Rotation Operators

Reflection and projection operators can be defined in a general sense by the way that they behave when applied repeatedly.

A *reflection operator* is defined as any operator R which acts as its own inverse, i.e. $R^2 = I$.

A. Seagar, *Application of Geometric Algebra to Electromagnetic Scattering*,
DOI 10.1007/978-981-10-0089-8_3

A *projection operator* is defined as any operator P which is idempotent. This means only the first of a sequence of ℓ identical operators can produce any effect. In general for $\ell > 1$, $P^\ell = P$, and in particular for $\ell = 2$, $P^2 = P$.

A *rotation operator* Θ is defined in general as the sequential application of two different reflection operators: $\Theta = R_2 R_1$.

In addition to these general definitions, it is useful to also define a pair of special projection operators, $P_1 = Q^+$ and $P_2 = Q^-$, with particular properties. This pair of projection operators is constructed from a single reflection operator $R = Q$ according to the relations:

$$\begin{cases} Q^+ = \frac{1}{2}(I + Q) \\ Q^- = \frac{1}{2}(I - Q) \end{cases} \tag{3.1}$$

The change in notation from R's and P's to Q's is made to emphasise these new projection operators as being both derived from a single reflection operator, and to emphasise they are a specially related pair rather than just two arbitrary and unrelated projections operators.

From the way they are constructed the projection operators Q^+ and Q^- are endowed with the following three properties:

$$\begin{cases} (Q^\pm)^2 = Q^\pm & \text{idempotent} \\ Q^\pm Q^\mp = 0 & \text{mutually exclusive} \\ Q^+ + Q^- = I & \text{complementary} \end{cases} \tag{3.2}$$

The first property ensures that Q^+ and Q^- are indeed both projection operators. The second property is useful for splitting fields into two complementary components. The third property is useful for reconstructing a field from its complementary components. Note that orthogonal components are indeed complementary, but complementary components need not be orthogonal (see Sect. 3.4.1).

The ability to split and reconstruct fields in this way is used here for many purposes, as in Sects. 3.2.1, 3.2.3, 3.4, 5.2 and Chap. 8.

3.2 Reflection and Projection

3.2.1 Vectors

In the Clifford framework vectors and fields can be split into a combination of components, normal and tangential to a specified surface Σ, by employing two projection operators such as described in Sect. 3.1. For Clifford algebra the specific form of the operators is defined in terms of a vector $u = u_n + u_t$ and the unit vector n normal to the surface Σ. The vector u is shown in Fig. 3.1, with u_n as the

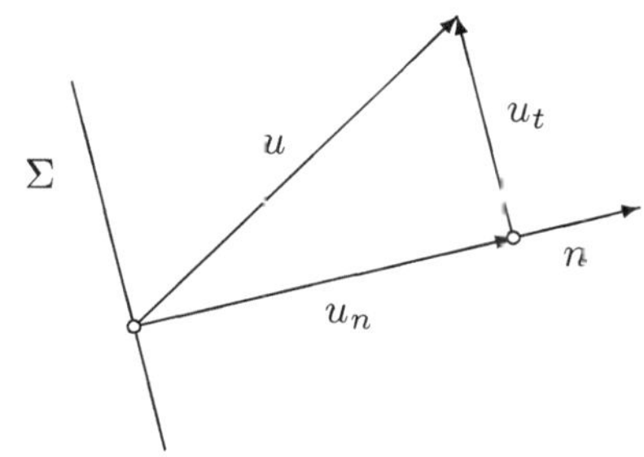

Fig. 3.1 Orthogonal components of vector u with respect to the unit vector n, normal to the surface Σ

component of u normal to Σ (parallel to n) and u_t as the component of u tangential to Σ (perpendicular to n).

The reflection operator is written as a product of three terms:

$$Ru = Qu = nun \tag{3.3}$$

and the pair of projection operators are written:

$$\begin{cases} P_1 u = Q^+ u = \frac{1}{2}(I + Q)u = \frac{1}{2}(u + nun) \\ P_2 u = Q^- u = \frac{1}{2}(I - Q)u = \frac{1}{2}(u - nun) \end{cases} \tag{3.4}$$

The action of the reflection operator can be determined by expanding the product 3.3 using the relationship 1.16 between Clifford multiplication and the vector dot and cross products. Expanding first the leftmost vector-vector product gives:

$$\begin{aligned} Qu = nun = [(n \times u)\sigma - (n \cdot u)]n &= (n \times u)\sigma n - (n \cdot u)n \\ &= \{(n \times u)n\}\sigma - (n \cdot u)n \end{aligned} \tag{3.5}$$

where the commutativity of σ with the primal vector units allows that $n\sigma = \sigma n$. Expanding now the remaining vector-vector product:

$$Qu = \{(n \times u) \times n\sigma - (n \times u) \cdot n\}\sigma - (n \cdot u)n \tag{3.6}$$

The second term in the braces is zero because $n \times u$ is perpendicular to n, and the first term simplifies after post multiplication since $\sigma^2 = 1$. Therefore:

$$Qu = (n \times u) \times n - (n \cdot u)n \tag{3.7}$$

The term $(n \cdot u)n$ is recognised from vector calculus as u_n, the component of u parallel to n (albeit with a minus sign) and the term $(n \times u) \times n$ is recognised as u_t, the component of u perpendicular to n. Rewriting Eq. 3.7 in terms of parallel and perpendicular components:

$$Q(u_t + u_n) = u_t - u_n \tag{3.8}$$

now shows that the action of the reflection operator is to reverse that component of vector u parallel to the unit vector n, while leaving the perpendicular component unchanged. The geometric interpretation in terms of Fig. 3.1 is that the vector u is reflected in the surface Σ for which n is the normal vector. In terms of the surface Σ, u_n is the normal component of vector u and u_t is the tangential component.

The action of the projection operators can be determined from that of the reflection operator:

$$\begin{cases} Q^{+}u = \frac{1}{2}(I + Q)u = \frac{1}{2}(u_t + u_n + u_t - u_n) = u_t = (n \times u) \times n \\ Q^{-}u = \frac{1}{2}(I - Q)u = \frac{1}{2}(u_t + u_n - u_t + u_n) = u_n = (n \cdot u)n \end{cases} \tag{3.9}$$

The projection operators $Q^{+}u$ and $Q^{-}u$ therefore split any Clifford vector u into normal and tangential components with respect to the surface Σ for which n is the normal vector.

3.2.2 Numerical Examples

A. Reflection

(a) Reflect the vector $u = 3e_1 + 3e_2$ in the y-axis.

- We need a unit vector perpendicular to the y-axis.
- Choose the unit vector $n_a = 1e_1 + 0e_2$.
- Then $R_a(u) = n_a u n_a = e_1(3e_1 + 3e_2)e_1 = (-3 + 3e_1e_2)e_1 = -3e_1 + 3e_2$
- See Fig. 3.2.

(b) Reflect the vector $u = 3e_1 + 3e_2$ in the line through the origin with slope 3/2.

- We need a unit vector perpendicular to the specified line.
- Choose the unit vector $n_b = \frac{1}{\sqrt{13}}(-3e_1 + 2e_2)$.

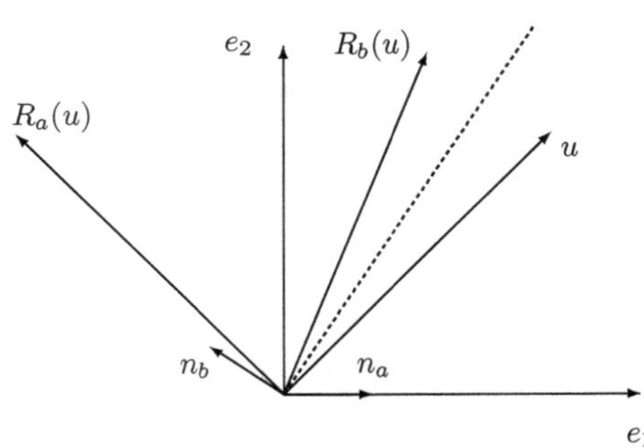

Fig. 3.2 Reflections $R_a = n_a u n_a$ and $R_b = n_b u n_b$

- Then $R_b(u) = n_b u n_b$
$= \frac{1}{\sqrt{13}}(-3e_1 + 2e_2)(3e_1 + 3e_2)\frac{1}{\sqrt{13}}(-3e_1 + 2e_2)$
$= \frac{1}{13}((9-6) + (-9-6)e_1e_2)(-3e_1 + 2e_2)$
$= \frac{1}{13}(3 - 15e_1e_2)(-3e_1 + 2e_2)$
$= \frac{1}{13}((-9+30)e_1 + (6+45)e_2)$
$= \frac{21}{13}e_1 + \frac{51}{13}e_2$
- See Fig. 3.2.

(c) Reflect the vector $u = 3e_1 + 4e_2 + 5e_3$ in the plane $x + z = 0$.

- We need a unit vector perpendicular to the plane.
- Choose the unit vector $n = \frac{1}{\sqrt{2}}(e_1 + e_3)$.
- Then $R(u) = nun$
$= \frac{1}{\sqrt{2}}(e_1 + e_3)(3e_1 + 4e_2 + 5e_3)\frac{1}{\sqrt{2}}(e_1 + e_3)$
$= \frac{1}{2}((-3-5) + 4e_1e_2 - 4e_2e_3 + (5-3)e_1e_3)(e_1 + e_3)$
$= \frac{1}{2}(-8 + 4e_1e_2 - 4e_2e_3 + 2e_1e_3)(e_1 + e_3)$
$= \frac{1}{2}(-8e_1 + 4e_2 - 4e_1e_2e_3 + 2e_3 - 8e_3 + 4e_1e_2e_3 + 4e_2 - 2e_1)$
$= \frac{1}{2}(-10e_1 + 8e_2 - 6e_3)$
$= -5e_1 + 4e_2 - 3e_3$

B. Projection

(a) Project the vector $u = 3e_1 + 3e_2$ onto the y-axis, and onto a line through the origin perpendicular to the y-axis.

- We need a unit vector perpendicular to the y-axis.
- Choose the unit vector $n_a = 1e_1 + 0e_2$.
- Then $P_1 = Q_a^+(u) = \frac{1}{2}[u + n_a u n_a]$
$= \frac{1}{2}[3e_1 + 3e_2 + e_1(3e_1 + 3e_2)e_1] = \frac{1}{2}[3e_1 + 3e_2 + (-3 + 3e_1e_2)e_1]$
$= \frac{1}{2}[3e_1 + 3e_2 - 3e_1 + 3e_2] \qquad = 3e_2$
- Also $P_2 = Q_a^-(u) = \frac{1}{2}[u - n_a u n_a]$
$= \frac{1}{2}[3e_1 + 3e_2 - e_1(3e_1 + 3e_2)e_1] = \frac{1}{2}[3e_1 + 3e_2 - (-3 + 3e_1e_2)e_1]$
$= \frac{1}{2}[3e_1 + 3e_2 + 3e_1 - 3e_2] \qquad = 3e_1$
- See Fig. 3.3.

(b) Project the vector $u = 3e_1 + 3e_2$ onto the line through the origin with slope 3/2, and onto a line through the origin perpendicular to the first one.

- We need a unit vector perpendicular to the specified line.
- Choose the unit vector $n_b = \frac{1}{\sqrt{13}}(-3e_1 + 2e_2)$.

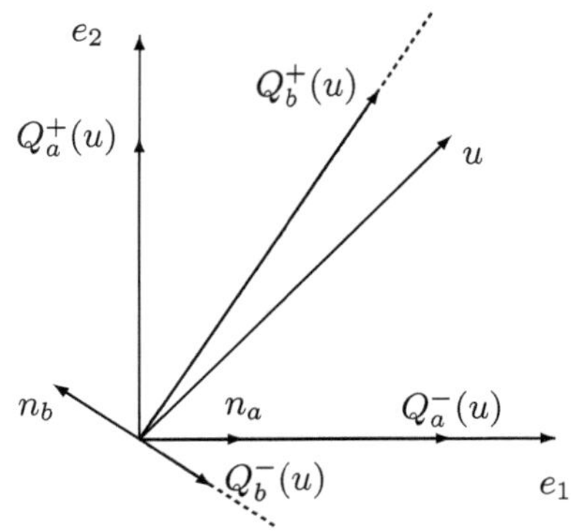

Fig. 3.3 Complementary projections $Q^{\pm}_{a,b} = \frac{1}{2}(I \pm Q_{a,b})$

- Then $P_1 = Q_b^+(u) = \frac{1}{2}[u + n_b u n_b]$
  $= \frac{1}{2}\left[3e_1 + 3e_2 + \frac{1}{\sqrt{13}}(-3e_1 + 2e_2)(3e_1 + 3e_2)\frac{1}{\sqrt{13}}(-3e_1 + 2e_2)\right]$
  $= \frac{1}{2}\left[3e_1 + 3e_2 + \frac{1}{13}((9-6) + (-9-6)e_1e_2)(-3e_1 + 2e_2)\right]$
  $= \frac{1}{2}\left[3e_1 + 3e_2 + \frac{1}{13}(3 - 15e_1e_2)(-3e_1 + 2e_2)\right]$
  $= \frac{1}{2}\left[3e_1 + 3e_2 + \frac{1}{13}((-9+30)e_1 + (6+45)e_2)\right]$
  $= \frac{1}{2}\left[3e_1 + 3e_2 + \frac{21}{13}e_1 + \frac{51}{13}e_2\right]$
  $= \frac{30}{13}e_1 + \frac{45}{13}e_2$
- Also $P_2 = Q_b^-(u) = \frac{1}{2}[u - n_b u n_b]$
  $= \frac{1}{2}\left[3e_1 + 3e_2 - \frac{1}{\sqrt{13}}(-3e_1 + 2e_2)(3e_1 + 3e_2)\frac{1}{\sqrt{13}}(-3e_1 + 2e_2)\right]$
  $= \frac{1}{2}\left[3e_1 + 3e_2 - \frac{1}{13}((9-6) + (-9-6)e_1e_2)(-3e_1 + 2e_2)\right]$
  $= \frac{1}{2}\left[3e_1 + 3e_2 - \frac{1}{13}(3 - 15e_1e_2)(-3e_1 + 2e_2)\right]$
  $= \frac{1}{2}\left[3e_1 + 3e_2 - \frac{1}{13}((-9+30)e_1 + (6+45)e_2)\right]$
  $= \frac{1}{2}\left[3e_1 + 3e_2 - \frac{21}{13}e_1 - \frac{51}{13}e_2\right]$
  $= \frac{9}{13}e_1 - \frac{6}{13}e_2$
- See Fig. 3.3.

(c) Project the vector $u = 3e_1 + 4e_2 + 5e_3$ onto the plane $x + z = 0$, and onto a line through the origin perpendicular to the plane.

- We need a unit vector perpendicular to the plane.
- Choose the unit vector $n = \frac{1}{\sqrt{2}}(e_1 + e_3)$.
- Then $P_1 = Q^+(u) = \frac{1}{2}[u + nun] = \frac{1}{2}[u + Q(u)]$
  $= \frac{1}{2}[(3e_1 + 4e_2 + 5e_3) + (-5e_1 + 4e_2 - 3e_3)] = \frac{1}{2}[-2e_1 + 8e_2 + 2e_3]$
  $= -e_1 + 4e_2 + e_3$
- Also $P_2 = Q^-(u) = \frac{1}{2}[u - nun] = \frac{1}{2}[u - Q(u)]$
  $= \frac{1}{2}[(3e_1 + 4e_2 + 5e_3) - (-5e_1 + 4e_2 - 3e_3)] = \frac{1}{2}[8e_1 + 8e_3] = 4e_1 + 4e_3$

3.2.3 Application to Bivector Fields

The electromagnetic field $\mathcal{F}$ is cast into the Clifford formalism as a bivector:

$$u = \mathcal{F} = \sqrt{\mu}\, H\sigma - i\sqrt{\epsilon}\, Ee_0 \tag{3.10}$$

where E and H are the electric and magnetic field vectors (see Sects. 1.7 and 5.1.2). Applying the reflection operator $R = Q$ gives:

$$\begin{aligned} Qu = nun \quad &= \sqrt{\mu}\, nH\sigma n - i\sqrt{\epsilon}\, nEe_0 n \\ = \sqrt{\mu}\,(nHn)\sigma + i\sqrt{\epsilon}\,(nEn)e_0 &= \sqrt{\mu}\,(QH)\sigma + i\sqrt{\epsilon}\,(QE)e_0 \end{aligned} \tag{3.11}$$

Both the magnetic and electric parts of the field are subject to the influence of the operator, with the electric part subject also to a change of sign.

Applying the projection operators $P_1 = Q^+$ and $P_2 = Q^-$ now gives:

$$\begin{aligned} Q^+u = \tfrac{1}{2}(I+Q)u &= \tfrac{1}{2}[\sqrt{\mu}\, H\sigma - i\sqrt{\epsilon}\, Ee_0 + \sqrt{\mu}\,(QH)\sigma + i\sqrt{\epsilon}\,(QE)e_0] \\ &= \sqrt{\mu}\,\tfrac{1}{2}(I+Q)H\sigma - i\sqrt{\epsilon}\,\tfrac{1}{2}(I-Q)Ee_0 \\ &= \sqrt{\mu}\,(Q^+H)\sigma - i\sqrt{\epsilon}\,(Q^-E)e_0 \\ &= \sqrt{\mu}\,(n \times H) \times n\sigma - i\sqrt{\epsilon}\,(n \cdot E)ne_0 \\ &= \sqrt{\mu}\, H_t\sigma - i\sqrt{\epsilon}\, E_n e_0 \end{aligned} \tag{3.12}$$

and:

$$\begin{aligned} Q^-u = \tfrac{1}{2}(I-Q)u &= \tfrac{1}{2}[\sqrt{\mu}\, H\sigma - i\sqrt{\epsilon}\, Ee_0 - \sqrt{\mu}\,(QH)\sigma - i\sqrt{\epsilon}\,(QE)e_0] \\ &= \sqrt{\mu}\,\tfrac{1}{2}(I-Q)H\sigma - i\sqrt{\epsilon}\,\tfrac{1}{2}(I+Q)Ee_0 \\ &= \sqrt{\mu}\,(Q^-H)\sigma - i\sqrt{\epsilon}\,(Q^+E)e_0 \\ &= \sqrt{\mu}\,(n \cdot H)n\sigma - i\sqrt{\epsilon}\,(n \times E) \times ne_0 \\ &= \sqrt{\mu}\, H_n\sigma - i\sqrt{\epsilon}\, E_t e_0 \end{aligned} \tag{3.13}$$

The projection operators Q^+ and Q^- therefore split the Clifford representation of the electromagnetic field into one part containing tangential magnetic field H_t paired with normal electric field E_n, and another part containing normal magnetic field H_n paired with tangential electric field E_t.

For some simple problems in electromagnetism the classical boundary conditions can be written directly (see Sects. 6.4.1 and 6.4.2) using either Q^-u (the normal component of the magnetic field and the tangential component of the electric field) or Q^+u (the tangential component of the magnetic field and the normal component of the electric field). In more complicated problems (see Sect. 6.4.3) it is necessary to use a combination of both Q^+ and Q^-, and also the S and T space-like and time-like projection operators as described in Sect. 5.2.

3.3 Rotation

A rotation operator Θ for any Clifford number u is written as a product of five terms:

$$\Theta u = R_2 R_1 u = n_2 n_1 u n_1 n_2 \tag{3.14}$$

where n_1 and n_2 are Clifford unit vectors, and $R_1 u = n_1 u n_1$ and $R_2 v = n_2 v n_2$ are two different reflection operators.

The geometric behaviour of the operator Θ is not immediately clear from Eq. 3.14. However, after a closer inspection (in Sect. 3.3.1) it becomes apparent that in general Θ indeed does behave as a rotation operator, and in particular that:

- the rotation is in the plane spanned by the two vectors n_1 and n_2,
- the angle of rotation θ is twice the smallest angle ϕ between n_1 and n_2, and
- the direction of rotation is from n_1 to n_2 (covering angle ϕ) and beyond (by another angle ϕ).

3.3.1 Vectors

The action of the rotation operator can be determined by expanding the product 3.14 using the relationship 1.16 between Clifford multiplication and the vector dot and cross products. Expanding first the left and rightmost vector-vector products, and then continuing with any remaining vector-vector products gives:

$$\begin{aligned}
\Theta u &= n_2 n_1 u n_1 n_2 \\
&= [-(n_2 \cdot n_1) + (n_2 \times n_1)\sigma] u [-(n_1 \cdot n_2) + (n_1 \times n_2)\sigma] \\
&= (n_2 \cdot n_1) u (n_1 \cdot n_2) - (n_2 \cdot n_1) u (n_1 \times n_2)\sigma - (n_2 \times n_1)\sigma u (n_1 \cdot n_2) \\
&\quad + (n_2 \times n_1)\sigma u (n_1 \times n_2)\sigma \\
&= (n_1 \cdot n_2) u (n_1 \cdot n_2) - (n_1 \cdot n_2) u (n_1 \times n_2)\sigma + (n_1 \times n_2) u (n_1 \cdot n_2)\sigma \\
&\quad - (n_1 \times n_2) u (n_1 \times n_2) \\
&= (n_1 \cdot n_2) u (n_1 \cdot n_2) - (n_1 \cdot n_2)[u(n_1 \times n_2) - (n_1 \times n_2) u]\sigma \\
&\quad - (n_1 \times n_2) u (n_1 \times n_2) \\
&= (n_1 \cdot n_2) u (n_1 \cdot n_2) - 2(n_1 \cdot n_2)[u \times (n_1 \times n_2)] \\
&\quad - (n_1 \times n_2) u (n_1 \times n_2)
\end{aligned} \tag{3.15}$$

In order to verify that Eq. 3.15 represents a rotation, consider now Fig. 3.4. The vector u is shown in a Cartesian coordinate system as the sum of two components:

$$u = v + w \tag{3.16}$$

where the v component is the projection of u onto the xy plane $z = 0$, and the w component is the projection of u onto the z-axis.

The Cartesian coordinate system has been set in space so that the plane $z = 0$ coincides with the plane spanned by the two unit vectors n_1 and n_2. Those unit vectors are not shown to avoid making the diagram too cluttered. However, regardless of their exact locations in the plane $z = 0$, their cross product stands perpendicular to the plane as shown along the z-axis.

The first term of Eq. 3.15 can be interpreted as a vector in the same direction as u but with length equal to that of u scaled by $\cos^2\phi$, where $\cos\phi = n_1 \cdot n_2$. This shorter version of u can alternatively be written as $u\cos^2\phi$ (as in Fig. 3.4) or as $(v + w)\cos^2\phi$.

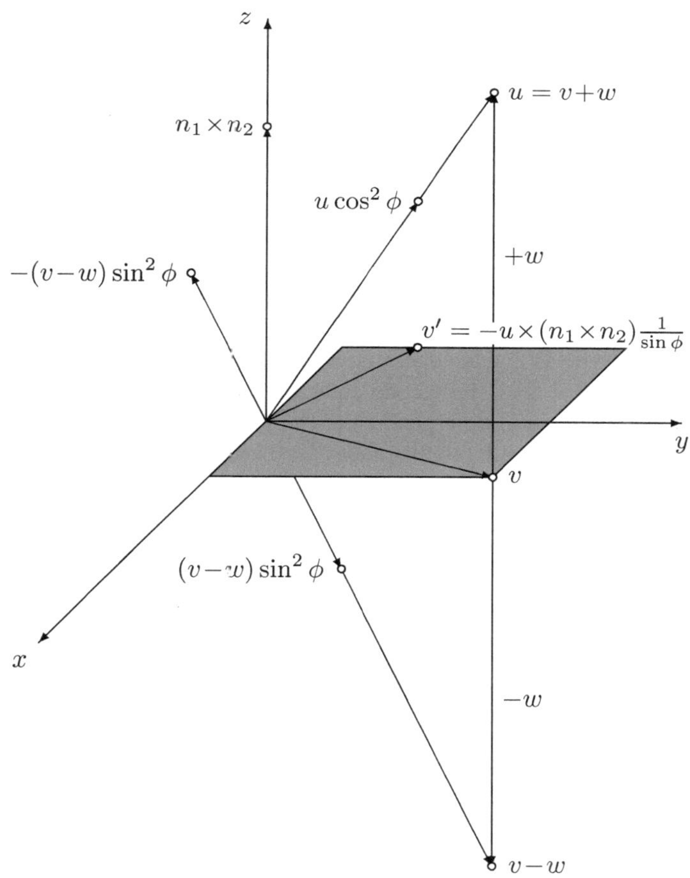

Fig. 3.4 Geometrical interpretation of rotation operator Θ

The second term of Eq. 3.15 can be simplified by noting that $u \times (n_1 \times n_2) = (v+w) \times (n_1 \times n_2)$. But w is parallel to $n_1 \times n_2$ and the cross product of two parallel vectors is zero. This leaves $u \times (n_1 \times n_2) = v \times (n_1 \times n_2)$, which represents a new vector in the plane $z = 0$, perpendicular to v, with length equal to that of v scaled by $\sin\phi = |n_1 \times n_2|$. In Fig. 3.4 this new vector has been rescaled and drawn as $v' = -\frac{1}{\sin\phi} u \times (n_1 \times n_2)$. The rescaling matches the length of v' with that of v, and chooses an orientation so that $v \times v'$ is in the same direction as $n_1 \times n_2$. Together the vectors v and v' serve as an orthogonal basis for the plane $z = 0$.

The third term of Eq. 3.15 represents a reversal of the component of vector u parallel to $n_1 \times n_2$, and at the same time (because $n_1 \times n_2$ is not a unit vector) a scaling of $\sin^2\phi$. The reversal of $w \to -w$ as shown in Fig. 3.4 has the effect of reflecting $u = v + w$ in the plane $z = 0$ giving an image vector $v - w$. Scaling by $\sin^2\phi$ gives $(v - w)\sin^2\phi$, and then scaling by -1 gives $-(v - w)\sin^2\phi$, both as shown.

Translating Eq. 3.15 in terms of the geometrical interpretation of Fig. 3.4 now gives:

$$\Theta u = (v + w)\cos^2\phi + 2v'\cos\phi\sin\phi - (v - w)\sin^2\phi \tag{3.17}$$

Projecting onto the plane $z = 0$ and onto the z-axis gives the same equation in two orthogonal components:

$$\begin{cases} \Theta v &= v\cos^2\phi + 2v'\cos\phi\sin\phi - v\sin^2\phi = v\cos 2\phi + v'\sin 2\phi \\ \Theta w &= w\cos^2\phi + w\sin^2\phi \qquad\qquad\qquad\quad = w \end{cases} \tag{3.18}$$

The first equation shows that the operator Θ rotates the component of u in the plane $z = 0$ by an angle $\theta = 2\phi$. The second equation shows that the operator Θ has no effect on any component of u perpendicular to the plane $z = 0$. As a consequence, all points on all planes parallel to the plane $z = 0$ are rotated by angle θ.

3.3.2 Numerical Examples

(a) Rotate the vector $u = 3e_1 + 3e_2$ through twice the smallest angle between the unit vectors $n_1 = 1e_1 + 0e_2$ and $n_2 = \frac{1}{\sqrt{13}}(-3e_1 + 2e_2)$, in the direction from n_1 to n_2.

- The angle between the two vectors is 146.31°.
- The rotation will be an angle of 2 × 146.31° = 292.62°.
- $\Theta(u) = R_2(R_1(u)) = n_2 n_1 u n_1 n_2$.
- First do inner reflection $v = R_1(u) = n_1 u n_1$
  $= e_1(3e_1 + 3e_2)e_1 = (-3 + 3e_1e_2)e_1 = -3e_1 + 3e_2$

Fig. 3.5 Rotation $\Theta = R_2 R_1$

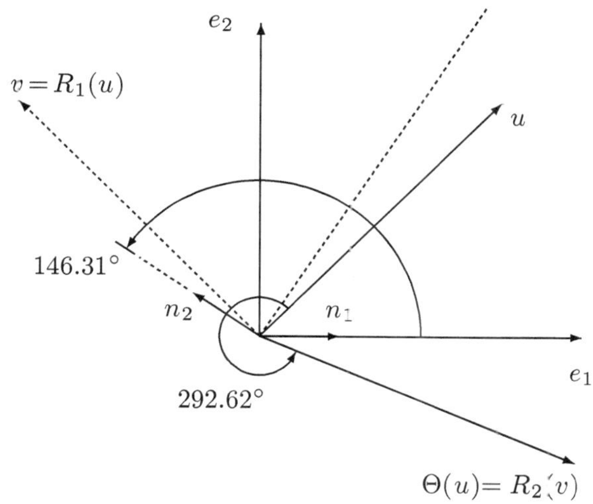

- Now do outer reflection $R_2(v) = n_2 v n_2$
  $= \frac{1}{\sqrt{13}}(-3e_1 + 2e_2)(-3e_1 + 3e_2)\frac{1}{\sqrt{13}}(-3e_1 + 2e_2)$
  $= \frac{1}{13}((-9 - 6) + (-9 + 6)e_1e_2)(-3e_1 + 2e_2)$
  $= \frac{1}{13}(-15 - 3e_1e_2)(-3e_1 + 2e_2)$
  $= \frac{1}{13}((45 + 6)e_1 + (-30 + 9)e_2)$
  $= \frac{51}{13}e_1 - \frac{21}{13}e_2$
- See Fig. 3.5.

(b) Rotate the vector $u = 3e_1 + 4e_2 + 5e_3$ by 90° around an axis through the origin perpendicular to the plane $x - z = 0$, in a direction so that the point $e_1 + e_3$ maps onto the positive y-axis (e_2 direction).

- We need two unit vectors in the plane separated by 45°.
- Choose the first unit vector $n_1 = \frac{1}{\sqrt{2}}(e_1 + e_3)$.
- Choose the second unit vector $n_2 = \frac{1}{\sqrt{2}}[\frac{1}{\sqrt{2}}(e_1 + e_3) + e_2] = \frac{1}{2}(e_1 + \sqrt{2}e_2 + e_3)$.
- $\Theta(u) = R_2(R_1(u)) = n_2 n_1 u n_1 n_2$.
- First do inner reflection $v = R_1(u) = n_1 u n_1$.
- Use result $v = -5e_1 + 4e_2 - 3e_3$ from example A(c) in Sect. 3.2.2.
- Now do outer reflection $R_2(v) = n_2 v n_2$
  $= \frac{1}{2}(e_1 + \sqrt{2}e_2 + e_3)(-5e_1 + 4e_2 - 3e_3) \times \frac{1}{2}(e_1 + \sqrt{2}e_2 + e_3)$
  $= \frac{1}{4}[(5 - 4\sqrt{2} + 3) + (4 + 5\sqrt{2})e_1e_2 - (4 + 3\sqrt{2})e_2e_3 + (5 - 3)e_1e_3]$
  $\times (e_1 + \sqrt{2}e_2 + e_3)$
  $= \frac{1}{4}[(8 - 4\sqrt{2}) + (4 + 5\sqrt{2})e_1e_2 - (4 + 3\sqrt{2})e_2e_3 + 2e_1e_3](e_1 + \sqrt{2}e_2 + e_3)$
  $= \frac{1}{4}[(8 - 4\sqrt{2} - 4\sqrt{2} - 10 - 2)e_1 + (8\sqrt{2} - 8 + 4 + 5\sqrt{2} + 4 + 3\sqrt{2})e_2$
  $+(8 - 4\sqrt{2} - 4\sqrt{2} - 6 + 2)e_3 + (4 + 5\sqrt{2} - 4 + 3\sqrt{2} - 2\sqrt{2})e_1e_2e_3]$
  $= -(1 + 2\sqrt{2})e_1 + 4\sqrt{2}e_2 + (1 - 2\sqrt{2})e_3$

3.3.3 Application to Bivector Fields

The electromagnetic field $\mathcal{F}$ is cast into the Clifford formalism, as for Eq. 3.10, in terms of a bivector $u = \mathcal{F} = \sqrt{\mu}H\sigma - i\sqrt{\epsilon}Ee_0$. Applying the rotation operator Θ gives:

$$\begin{aligned}\Theta u &= n_2 n_1 u n_1 n_2 \\ &= n_2[\sqrt{\mu}\, n_1 H\sigma n_1 - i\sqrt{\epsilon}\, n_1 E e_0 n_1] n_2 \\ &= n_2[\sqrt{\mu}\,(n_1 H n_1)\sigma + i\sqrt{\epsilon}\,(n_1 E n_1) e_0] n_2 \\ &= \sqrt{\mu}\, n_2 (n_1 H n_1)\sigma n_2 + i\sqrt{\epsilon}\, n_2 (n_1 E n_1) e_0 n_2 \\ &= \sqrt{\mu}\,(n_2 n_1 H n_1 n_2)\sigma - i\sqrt{\epsilon}\,(n_2 n_1 E n_1 n_2) e_0 \\ &= \sqrt{\mu}\,(\Theta H)\sigma - i\sqrt{\epsilon}\,(\Theta E) e_0 \end{aligned} \tag{3.19}$$

Both the magnetic and electric parts of the field are subject to the influence of the operator. Note that the same effect can be achieved by first applying the rotation operator to the individual parts of the field and then combining the results as an electromagnetic bivector, or by assembling the field as a first step and then applying the rotation to that bivector field.

3.4 Coordinate Systems

Coordinate systems provide a representation of vectors (or functions) in terms of several independent components which upon summation reconstruct the original entity. For example, any vector can be represented in relation to a surface in terms of its normal and tangential components, and any function of one real variable can be represented in terms of its even and odd components. In both cases these components can be created using a pair of suitable projection operators.

For problems in electromagnetism the behaviour of the field on a boundary can be portrayed by certain projection operators P^+, P^- (Hardy projection operators, see Chap. 8), and the boundary conditions can be formulated by certain other projection operators Q^+, Q^- (boundary data projection operators, see Chap. 6). General tools for working directly with these operators help make much easier the task of formulating electromagnetic field problems in a clear, concise and efficacious manner.

The tools for working with the boundary data projection operators inhabit normal Euclidean space. However, the tools for working with the Hardy projection operators inhabit the domain of Banach space. This space is a complete linear functional space which has a norm, but not a scalar product. The equivalent geometric space which operates with vectors rather than functions is the space of affine geometry, as shown in Table 3.1. In affine geometries and Banach spaces the concept of parallelism applies but the concept of perpendicularity does not. In Euclidean geometries and Hilbert spaces the concepts of parallelism and perpendicularity both apply. Many of the figures in this book are drawn for Banach space using affine geometry, but because they are constructed in a Euclidean medium it appears that some of the angles may be right angles. They aren't. There are no right angles, even when they visually appear

Table 3.1 Equivalent geometric and functional spaces

	Nc scalar product	Scalar product
Vectors	Affine geometry	Euclidean geometry
Functions	Banach space	Hilbert space

that way. There is nothing perpendicular to anything else. There are really no angles at all. These figures can be subject to an arbitrary shear operation to apparently destroy the constructs that look like right angles, and still remain valid.

3.4.1 Affine Geometry

In a two-dimensional space of affine geometry, vectors $\binom{x}{y} = \binom{a}{b}$ can be represented in terms of two components:

$$\binom{x}{0} = \begin{pmatrix}1 & 0\\ 0 & 0\end{pmatrix}\binom{a}{b} \quad \text{and} \quad \binom{0}{y} = \begin{pmatrix}0 & 0\\ 0 & 1\end{pmatrix}\binom{a}{b} \tag{3.20}$$

The matrices $A^+ = \begin{pmatrix}1 & 0\\ 0 & 0\end{pmatrix} = \frac{1}{2}\left[\begin{pmatrix}1 & 0\\ 0 & 1\end{pmatrix} + \begin{pmatrix}1 & 0\\ 0 & -1\end{pmatrix}\right]$ and $A^- = \begin{pmatrix}0 & 0\\ 0 & 1\end{pmatrix} = \frac{1}{2}\left[\begin{pmatrix}1 & 0\\ 0 & 1\end{pmatrix} - \begin{pmatrix}1 & 0\\ 0 & -1\end{pmatrix}\right]$ are projection operators created as in Sect. 3.1 from a single reflection operator $A = \begin{pmatrix}1 & 0\\ 0 & -1\end{pmatrix}$.

The projection operator A^+ produces all points on the $y = 0$ axis $OA : \binom{x}{y} = \binom{a}{0}$ as depicted in Fig. 3.6. The projection operator A^- produces all points on the $x = 0$ axis $OA' : \binom{x}{y} = \binom{0}{b}$. The origin $\binom{0}{0}$ lies at the intersection of both axes. Because the two projections are complementary ($A^+ + A^- = I$) all points $\binom{x}{y} = \binom{a}{b}$ can be reconstructed from their two components.

The projections A^+ and A^- properly define two Cartesian coordinate axes $\binom{x}{0}$ and $\binom{0}{y}$. The components of any point $\binom{x}{y} = \binom{a}{b}$ can be obtained by projecting it onto the two axes as in Eq. 3.20.

Although all projection operators do produce axes, not all pairs of axes can serve in the role of coordinate axes; only those produced from complementary projection operators. Consider for example, the two projection operators $P_1 = \frac{1}{2}\left[\begin{pmatrix}1 & 0\\ 0 & 1\end{pmatrix} + \begin{pmatrix}1 & 0\\ 0 & -1\end{pmatrix}\right]$ and $P_2 = \frac{1}{2}\left[\begin{pmatrix}1 & 0\\ 0 & 1\end{pmatrix} - \begin{pmatrix}0 & 1\\ 1 & 0\end{pmatrix}\right]$. The projection operator $P_1 \binom{a}{b}$ produces all points on the

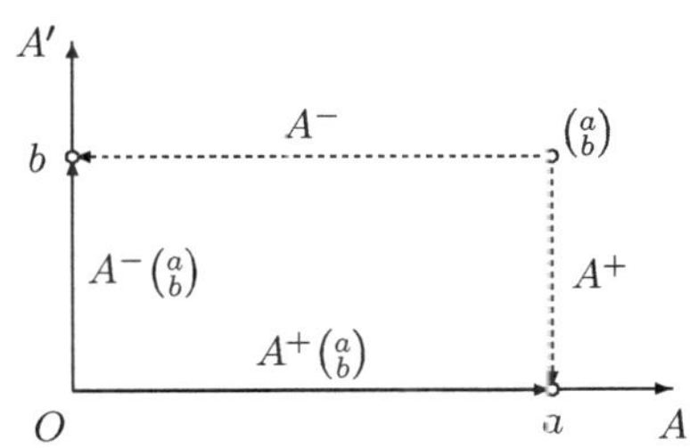

Fig. 3.6 Orthogonal coordinate axes OA, OA' defined by projections $A^{\pm}$

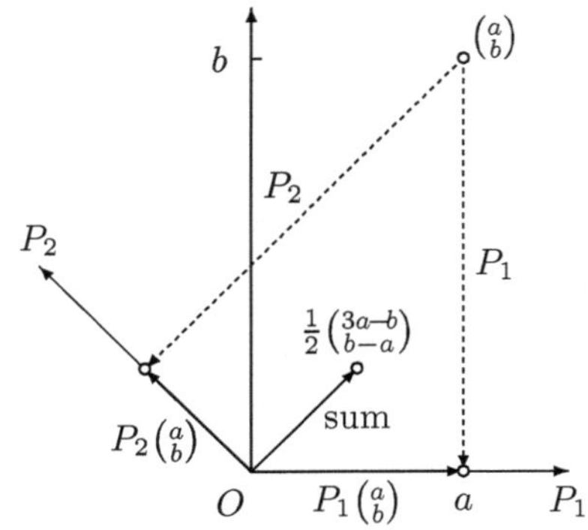

Fig. 3.7 Axes OP_1, OP_2 defined by projections P_1, P_2 not suitable as coordinate axes

$y = 0$ axis $OP_1 : \binom{x}{y} = \binom{a}{0}$ as shown in Fig. 3.7. $P_2 \binom{a}{b}$ produces all points on the $y = -x$ axis $OP_2 : \binom{x}{y} = \frac{1}{2}\binom{a-b}{b-a}$ through the origin at an apparent slope of $-1:1$. Summing the two components of point $\binom{a}{b}$ along the axes fails to reconstruct the original point:

$$P_1 \binom{a}{b} + P_2 \binom{a}{b} = \binom{a}{0} + \frac{1}{2}\binom{a-b}{b-a} \neq \binom{a}{b} \tag{3.21}$$

Axes are identified as *coordinate axes* if the projection operators which produce each axis map all points on that axis back onto themselves ($A^{\pm}A^{\pm} = A^{\pm}$), and map all points on other axes to the origin ($A^{\pm}A^{\mp} = 0$). Axes are identified as unsuitable for coordinate axes if the projections operators which produce each axis map points on any other axes to anywhere other than the origin, for example $P_1P_2 \binom{a}{b} = \frac{1}{2}\begin{pmatrix}1 & -1\\ 0 & 0\end{pmatrix}\binom{a}{b} \neq \binom{0}{0}$.

Axes which do serve as pairs of coordinate axes are identified here, as for *OA* and *OA*′ in Fig. 3.6, by primed and unprimed versions of the same label.

Although it is common in Euclidean geometry to adopt coordinate axes which are orthogonal, affine geometry has no sense of othorgonality so it isn't possible. Fortunately, orthogonality between a pair of axes is not necessary for them to serve as coordinate axes. Consider for example the two complementary projection operators $P^+ = \frac{1}{2}\left[\begin{pmatrix}1 & 0\\ 0 & 1\end{pmatrix} + \begin{pmatrix}1 & -2\\ 0 & -1\end{pmatrix}\right]$ and $P^- = \frac{1}{2}\left[\begin{pmatrix}1 & 0\\ 0 & 1\end{pmatrix} - \begin{pmatrix}1 & -2\\ 0 & -1\end{pmatrix}\right]$. The projection operator P^+ produces all points on the $y = 0$ axis $OP : \binom{x}{y} = \binom{a-b}{0}$ as depicted in Fig. 3.8. P^- produces all points on the $y = x$ axis $OP' : \binom{x}{y} = \binom{b}{b}$ through the origin at an apparent slope of $+1:1$. The *OP* and *OP*′ axes are clearly not orthogonal. However summing the two components of point $\binom{a}{b}$ along the axes certainly does reconstruct the original point:

$$P^+ \binom{a}{b} + P^- \binom{a}{b} = \binom{a-b}{0} + \binom{b}{b} = \binom{a}{b} \tag{3.22}$$

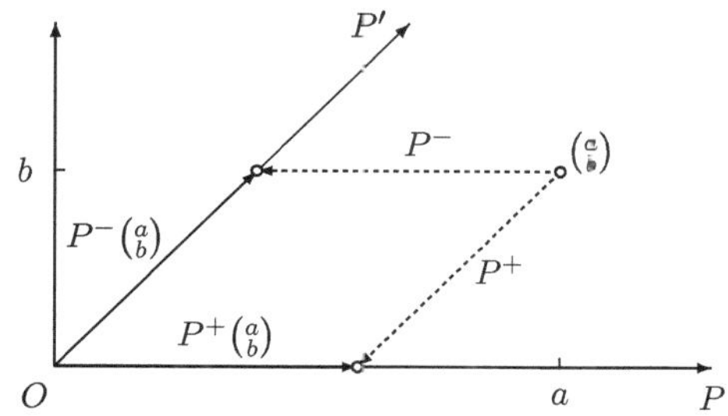

Fig. 3.8 Non-orthogonal OP, OP' coordinate axes defined by projections $P^{\pm}$

3.4.2 Banach Space

In a two-dimensional Banach space, functions f can be represented in terms of two components:

$$f_A = A^+ f \qquad \text{and} \qquad f_{A'} = A^- f \tag{3.23}$$

The operators $A^+ = \frac{1}{2}(I+A)$ and $A^- = \frac{1}{2}(I-A)$ are projection operators created as in Sect. 3.1 from a single reflection operator A. The projection operator A^+ produces all functions f_A on the OA axis, and the projection operator A^- produces all functions $f_{A'}$ on the OA' axis. The two axes can be interpreted as straight lines because all points on each axis map by projection onto a single point. For the OA axis $A^- f_A = A^- A^+ f = 0$, and for the OA' axis $A^+ f_{A'} = A^+ A^- f = 0$. Were the axes not straight that would not happen. The axes intersect at the origin O and, because the two projections are complementary ($A^+ + A^- = I$), all functions f can be reconstructed from their two components.

The geometry is drawn either as in Fig. 3.6, or as in Fig. 3.8 with P replaced by A, and the vector $\binom{a}{b}$ replaced by the function f. In terms of affine geometry, there is no distinction between the two figures because there is no concept of orthogonality and no measure of angle. So the visually apparent angles between certain lines in both of the figures cannot be 90° or anything else. The reason these constructs appear to be right angles is because we are forced to draw these affine diagrams within the Euclidean space that we, but not this particular geometry, inhabit.

The complementary projections A^+ and A^- properly define two Cartesian coordinate axes OA and OA' in Banach space.

The components of any function f can be obtained by projecting it onto the two axes as in Eq. 3.23.

3.4.3 Transformations Between Coordinate Systems

Figure 3.9 shows two Cartesian systems in Banach space: the P-system with axes OP and OP', and the Q-system with axes OQ and OQ'. Both systems share the same origin O.

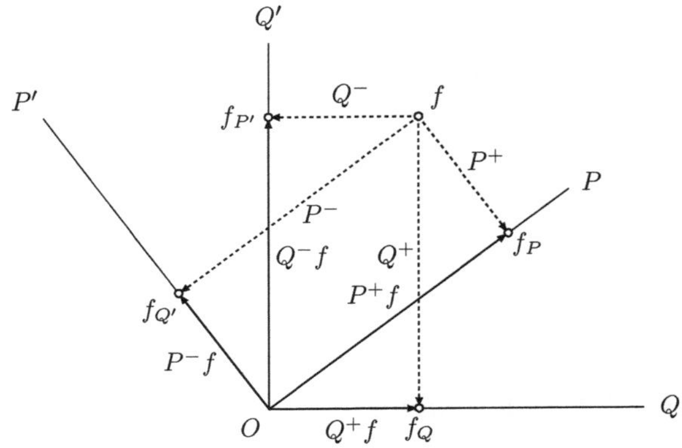

Fig. 3.9 Components of function f in two different Cartesian systems

The P-system is based on the Cauchy integral C_Σ and on the pair of Hardy projection operators P^+ and P^-, as described in Chaps. 7 and 8. The Q-system is based on the boundary data reflection operator Q and corresponding projection operators Q^+ and Q^-, as described in Chap. 6.

A function f can be represented in terms of Cartesian components in either the P-system, or in the Q-system:

$$\begin{cases} f = f_P + f_{P'} = P^+ f + P^- f \\ f = f_Q + f_{Q'} = Q^+ f + Q^- f \end{cases} \tag{3.24}$$

The conversions from components f_P, $f_{P'}$ to components f_Q, $f_{Q'}$ are straightforward projections:

$$\begin{cases} f_Q = Q^+ f = Q^+(f_P + f_{P'}) = Q^+ f_P + Q^+ f_{P'} \\ f_{Q'} = Q^- f = Q^-(f_P + f_{P'}) = Q^- f_P + Q^- f_{P'} \end{cases} \tag{3.25}$$

or:

$$\begin{pmatrix} f_Q \\ f_{Q'} \end{pmatrix} = \begin{pmatrix} Q^+ & Q^+ \\ Q^- & Q^- \end{pmatrix} \begin{pmatrix} f_P \\ f_{P'} \end{pmatrix} \tag{3.26}$$

Conversions from components f_Q, $f_{Q'}$ to components f_P, $f_{P'}$ are equally straightforward:

$$\begin{cases} f_P = P^+ f = P^+(f_Q + f_{Q'}) = P^+ f_Q + P^+ f_{Q'} \\ f_{P'} = P^- f = P^-(f_Q + f_{Q'}) = P^- f_Q + P^- f_{Q'} \end{cases} \tag{3.27}$$

or:

$$\begin{pmatrix} f_P \\ f_{P'} \end{pmatrix} = \begin{pmatrix} P^+ & P^+ \\ P^- & P^- \end{pmatrix} \begin{pmatrix} f_Q \\ f_{Q'} \end{pmatrix} \tag{3.28}$$

Note that the conversions are projections rather than the rotations that would be expected in a Euclidean space. Banach space lacks a scalar product, which means that angles cannot be defined and rotations are therefore not supported.

3.4.4 Multi-Dimensional Axes

Axes defined by projection can represent single or multi-dimensional spaces. In Fig. 3.10a the axes OX, OY and OZ defined by projections:

$$P_x = \begin{pmatrix} 1\,0\,0 \\ 0\,0\,0 \\ 0\,0\,0 \end{pmatrix}, \qquad P_y = \begin{pmatrix} 0\,0\,0 \\ 0\,1\,0 \\ 0\,0\,0 \end{pmatrix}, \qquad P_z = \begin{pmatrix} 0\,0\,0 \\ 0\,0\,0 \\ 0\,0\,1 \end{pmatrix} \tag{3.29}$$

each represent a single dimension in a three-dimensional space. Points on any one of the axes are associated with single scalar values of coordinates.

The same three-dimensional space can be represented more compactly as in Fig. 3.10b by the two axes $O\overline{XY}$ and OZ as defined by the projections:

$$P_{xy} = \begin{pmatrix} 1\,0\,0 \\ 0\,1\,0 \\ 0\,0\,0 \end{pmatrix}, \qquad P_z = \begin{pmatrix} 0\,0\,0 \\ 0\,0\,0 \\ 0\,0\,1 \end{pmatrix} \tag{3.30}$$

In the latter representation the points on the axis $O\overline{XY}$ are associated with values of coordinates which are pairs of numbers rather than single numbers. Together the two axes span three dimensions.

For the case of electromagnetism, points on the coordinate axes in Banach space defined by Hardy and boundary data projections are associated with values which are N-tuples. Each component of these N-tuples takes the (Clifford) value of the field at one of N locations representing the boundary enclosing a region. That allows multi-dimensional figures in Banach space to be drawn compactly on the two-dimensional pages of this book.

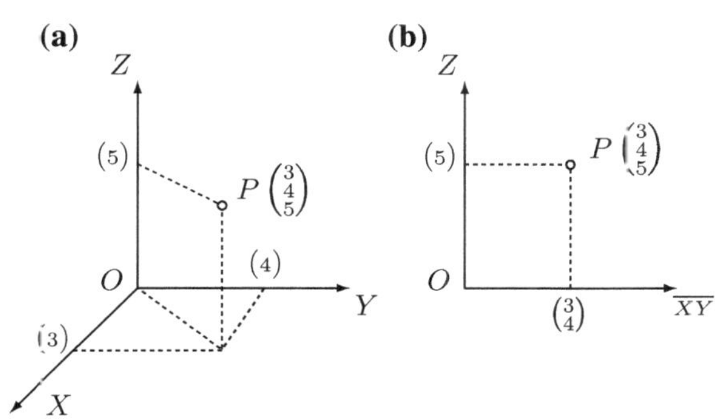

Fig. 3.10 Axes defined by projection (**a**) one-dimensional only (**b**) one and two-dimensional

3.5 Intersection by Projection

Physical problems of a linear nature often lead to a formulation in terms of a linear system, such as a matrix equation $Ax = b$, for which the coefficients A and data b are known, and which must be inverted in some way to determine the unknown x.

Electromagnetism is a linear problem in that the solution for two sources is the same as the sum of solutions for each of these sources separately, and some kind of inversion is required when finding the field $x = \mathcal{F}$ from the sources $b = \mathcal{S}$.

The solution of a linear system in m variables can be interpreted geometrically as the intersection of m non-parallel hyperplanes at a single point x, and can usually be handled by any of the conventional direct or iterative techniques (e.g. Gaussian elimination, Gauss-Seidel iteration, or variants and improvements thereof). These techniques are general and take no specific account of the particular structure of any given problem.

The problem of interest here is that of the scattering of electromagnetic radiation from electrically non-conducting and conducting objects. For certain types of materials the formulation leads to two equations, each (as in Sects. 9.1 and 9.2) giving one vector as the projection of another.

The two equations represent two intersecting straight lines in Banach space. One of the lines represents all electromagnetic fields on the surface of an object which are consistent with Maxwell's equations and the other line represents all electromagnetic fields which are consistent with the boundary conditions. The solution for a particular scattering problem with particular boundary conditions is found at the intersection of the lines. Direct application of matrix methods does not work because projection operators are always singular.

In order to solve the electromagnetic scattering problem as posed here, it therefore makes sense to seek a solution as the intersection of lines, or planes, or hyperplanes directly in terms of singular but known projection operators $P_1, P_2 \ldots P_m$, rather than in terms of a non-singular but unknown linear system $Ax = b$.

The method described in Sect. 3.5.1 employs successive projections onto constraints, first used by Kaczmarz [23, 24] for orthogonal projections, together with Richardson extrapolation [25, 26] to sum an infinite geometric series of vectors. Here the projections are not orthogonal, but the approach is essentially the same.

3.5.1 Formulation

Consider the case shown in Fig. 3.11 with two lines intersecting at a point x. A point x_0 on one of the lines is projected onto the other line according to projection P_1, and then from there back onto the original line at point x_1 using projection P_2:

$$x_1 = P_2 P_1 x_0 \tag{3.31}$$

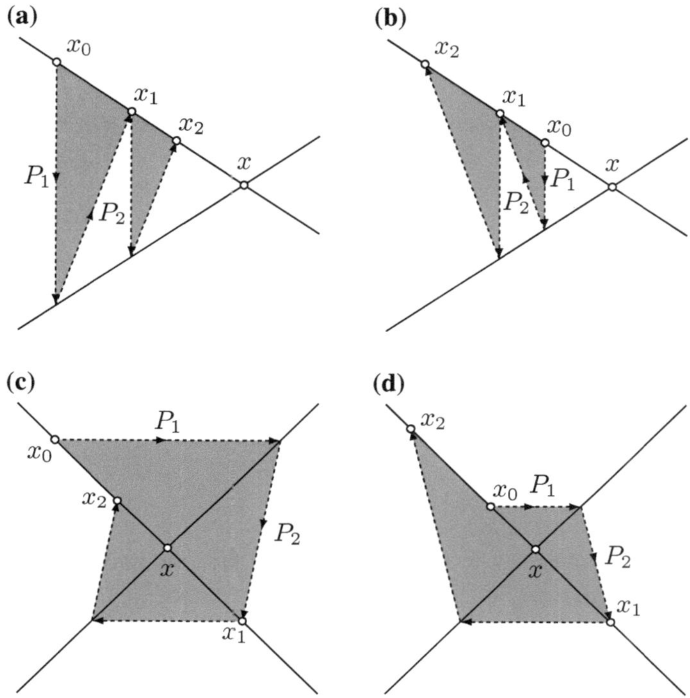

Fig. 3.11 Intersection of lines by projection (**a**) $0 < \xi < 1$, (**b**) $1 < \xi$, (**c**) $-1 \leq \xi < 0$, (**d**) $\xi \leq -1$

It is important that the two projections are independent, i.e. $P_2 P_1 \neq I$. Repeating the process gives another point x_2:

$$x_2 = P_2 P_1 x_1 \tag{3.32}$$

The two shaded triangles in each part of the figure are similar. Scaling the figures by a factor ξ, according to the ratio of vectors $(x_2 - x_1) : (x_1 - x_0) = \xi : 1$, maps the larger shaded triangle into the smaller one for Fig. 3.11a, c, and maps the smaller triangle into the larger one for Fig. 3.11b, d. In Fig. 3.11a, b the value of ξ is positive, whereas in Fig. 3.11c, d it is negative.

For Fig. 3.11a the intersection is found by adding to vector x_0, the vectors $x_1 - x_0$, $x_2 - x_1$, $x_3 - x_2$, and so on. All of the vectors added from $x_2 - x_1$ and beyond stand in the ratio, each to the previous, of $\xi : 1$. The point x can therefore be found by writing the infinite sum of vectors as an infinite geometric series:

$$x = x_0 + (x_1 - x_0)[1 + \xi + \xi^2 + \cdots] \tag{3.33}$$

The same result holds for Fig. 3.11b, c, d without modification. Scaling Eq. 3.33 by ξ and subtracting from the original eliminates the infinite series and gives:

$$\begin{aligned}(1-\xi)x &= (1-\xi)x_0 + (x_1 - x_0)\\ x &= x_0 + \nu(x_1 - x_0)\end{aligned} \tag{3.34}$$

where the value of ν is chosen such that $\nu(1-\xi)x = x$ for any value of x. The relationship:

$$\begin{aligned}(1-\xi)(x_1 - x_0) = (x_1 - x_0) - (x_2 - x_1) &= (2x_1 - x_0 - x_2)\\ (x_1 - x_0) &= \nu(2x_1 - x_0 - x_2)\end{aligned} \tag{3.35}$$

between the three vectors x_0, x_1, x_2 provides a means to obtain the value of ν.

In two-dimensional Euclidean space, where the vector x represents a point in the real plane, the scalars ξ and ν are real numbers. In this situation, the norm $\nu\|z\|$ induced by the scalar product $(\nu z, z) = \nu(z, z)$ can be used to convert the vector Eq. 3.35 into the scalar equation:

$$(x_1 - x_0,\ 2x_1 - x_0 - x_2) = \nu(2x_1 - x_0 - x_2,\ 2x_1 - x_0 - x_2) \tag{3.36}$$

The value of ν is obtained by scalar division and then the point of intersection is found directly from Eq. 3.34.

The value of ν is well defined except in the case that the two lines are parallel. Then $\xi = 1$ and it is not possible to find a proper value of ν. Excepting that case, the intersection can be found using Eq. 3.34. Any starting point can be used, projecting first onto x_0 using P_2, and then onto x_1 and x_2.

In m Euclidean dimensions exactly the same method holds, with x as the point of intersection between m hyperplanes. In this case point x_0 on any hyperplane is projected onto all hyperplanes in any chosen order before returning to the original hyperplane:

$$x_1 = P_m \ldots P_2 P_1 x_0 \tag{3.37}$$

Point x_2 is calculated from x_1 using the previously chosen order of projections, then ν from x_0, x_1, x_2, and the intersection x from Eq. 3.34. Point x_0 can be obtained from any arbitrary starting point by projecting onto the first hyperplane using projection P_m. If there is no particular reason for choosing one starting point as opposed to another, the origin can be used.

For the case of electromagnetism, although the problem can be cast in a diagram drawn in a two-dimensional affine space, the solution is less straightforward than for either two or m-dimensional Euclidean space. The reason is that the axes in the diagram are themselves both N-dimensional, representing the electromagnetic field at N points on the surface of an object. Mapping from the vectorial difference between one pair of N-dimensional points x_0, x_1 into the difference between another pair x_1, x_2, along an axis embedded within the space, is achieved by scaling not with a real number but with an $N \times N$ matrix. Equation 3.35 is then a matrix-vector equation with an $N \times N$ matrix ν and N-dimensional vectors x_j. Simple real scalar division is not sufficient to recover the value of ν except in the special case where it is diagonal. Normally that happens only in trivial cases, and in general a sequence of $N+1$ (rather than $1+1=2$) pairs of projections are required to build from Eq. 3.35 a square matrix-matrix equation. Solving that equation by matrix elimination then gives the value for the matrix ν.

3.5.2 Remarks

The method described here relies solely on the ability to calculate the projection of points onto hyperplanes. The projections can be in any direction, as long as the whole set is independent. If the problem is such that the projections can be chosen freely, natural choices would be along each of the coordinate axes, or in the direction of the normal of the next hyperplane. In the case of electromagnetism the projections arise as part of the formulation of the problem.

If the projections are chosen along coordinate axes, the resulting method is effectively that of Gauss-Seidel. Normally with Gauss-Seidel it is anticipated that the algorithm reaches the solution before point x_2 is calculated, so Eq. 3.34 is not used. Convergence with the Gauss-Seidel iteration is not guaranteed, and the iteration diverges in cases such as Fig. 3.11b, d. Equation 3.34 guarantees a solution in all cases where the projections are independent, except where there is none ($\xi = 1$).

An alternative choice of projections is in the direction of the normal to the next hyperplane. The method then takes into account the structure of the problem itself and, unlike Gauss-Seidel, is independent of the structure of the coordinate system in which the problem is expressed. However, if there is no unique solution projections in the direction of the normal violate the condition that all projections be independent. In that case Eq. 3.34 is not valid, should not be used, and if used will give a spurious result even when $\xi \neq 1$.

For the case of electromagnetism, in the spirit of Gauss-Seidel, it is anticipated that the iterated projections converge before point x_N is calculated. Convergence and speed of convergence are governed by the properties of the particular projections involved [27, 28], these being the boundary data projection operators $Q^{\pm}$ (see Chap. 6) and the Hardy projection operators $P^{\pm}$ (see Chap. 8). Numerical values of the convergence for certain test cases are given in Chap. 10.

3.6 Exercises

Q1. Show that a projection operator Q^+ constructed from a reflection operator Q as in Eq. 3.1 is idempotent.

A1. $(Q^+)^2 = \frac{1}{4}(I+Q)(I+Q) = \frac{1}{4}(I+Q+Q+Q^2) = \frac{1}{4}(I+Q+Q+I) = \frac{1}{2}(I+Q) = Q^+$.

Q2. Show that a pair of projection operators Q^+ and Q^- constructed from a reflection operator Q as in Eq. 3.1 are mutually exclusive.

A2. $(Q^+)(Q^-) = \frac{1}{4}(I+Q)(I-Q) = \frac{1}{4}(I-Q+Q-Q^2) = \frac{1}{4}(I-I) = 0$. Also $(Q^-)(Q^+) = 0$.

Q3. Show that a pair of projection operators Q^+ and Q^- constructed from a reflection operator Q as in Eq. 3.1 are complementary.

A3. $Q^+ + Q^- = \frac{1}{2}(I+Q) + \frac{1}{2}(I-Q) = I$.

Q4. Show that the rotation operator has no effect on scalars.

A4. For any scalar $u \in \bigwedge^0$, $\Theta u = n_2 n_1 u n_1 n_2 = n_2 n_1 n_1 n_2 u = -n_2 n_2 u = u$.

Q5. Show that the rotation operator has a distributive property over sums of Clifford numbers.

A5. For any Clifford numbers u and v, $\Theta(u+v) = n_2 n_1 (u+v) n_1 n_2 = n_2 n_1 u n_1 n_2 + n_2 n_1 v n_1 n_2 = \Theta u + \Theta v$.

Q6. Show that the rotation operator has a distributive property over products of Clifford numbers.

A6. For any Clifford numbers u and v, $\Theta(uv) = n_2 n_1 (uv) n_1 n_2 = (n_2 n_1 u)(v n_1 n_2)$ $= (n_2 n_1 u)(n_1 n_2 n_2 n_1)(v n_1 n_2) = (n_2 n_1 u n_1 n_2)(n_2 n_1 v n_1 n_2) = \Theta u \Theta v$.

Q7. Verify that the operators A^+ and A^- in Sect. 3.4.1 are projection operators, and that the operator $A^+ - A^-$ is a reflection operator.

Q8. Verify that the operators P_1 and P_2 in Sect. 3.4.1 are projection operators, but that the operator $P_1 - P_2$ is not a reflection operator.

Q9. Show that the sequence of two projection operators $P_2 P_1$ in Sect. 3.4.1 fails to map all points to the origin.

Q10. Show that Eq. 3.33 holds for the cases of Fig. 3.11b, c.

Q11. By choosing the projections $P_1, P_2 \ldots P_m$ in Eq. 3.37 along the coordinate axes show that the method described mimics the Gauss-Seidel iteration.

Q12. Design and implement a numerical procedure to solve a linear system of equations with real coefficients using orthogonal projections onto hyperplanes with the geometrical method described in Sects. 3.5 and 3.5.1.

Q13. Design and implement a numerical procedure to solve a linear system of equations with real coefficients using projections along Cartesian coordinate axes with the geometrical method described in Sects. 3.5 and 3.5.1.

Chapter 4
Space and Time

Maxwell's equations were developed before Einstein introduced space–time in his theory of special relativity. Nevertheless, results derived by solving Maxwell's equations must properly accommodate Einstein's theory in order to conform with the physical phenomena of electromagnetism observed in the static and moving real world. Only in that way can applications be designed and built to operate according to intent.

The need for that conformance puts a constraint on the way the derivative in space–time is mathematically constructed. In the Heaviside form [14][1] of Maxwell's equations the derivative is calculated using a set of three separate partial differential operators: two spatial (divergence and curl) and one temporal ($\frac{\partial}{\partial t}$). In the Clifford formulation all of these partial differential operators are combined into a single, complete differential operator $\mathcal{D}$. However, the need to meet Einstein's theory and observed physical behaviour means that only one of several possible mathematical combinations makes any physical sense.

In the sections following the Minkowskian geometry of Einstein's space–time is introduced in the Clifford formalism by way of the Lorentz transformation. A generic Clifford–Minkowskian differential operator is constructed and then constrained so as to properly accommodate Einstein's theory of special relativity. The result is verified by calculating the electromagnetic field observed in a moving frame from the field observed in a fixed frame.

---

[1] Sections 51 and 66.

A. Seagar, *Application of Geometric Algebra to Electromagnetic Scattering*,
DOI 10.1007/978-981-10-0089-8_4

4.1 Points in Space–Time

The space–time of special (not general) relativity uses Minkowskian geometry with one basis vector X_0 to establish a coordinate x_0 of time, and three basis vectors X_1, X_2, X_3 to establish three Cartesian coordinates x_1, x_2, x_3 of space. Any point X in Minkowskian space–time can be represented in terms of its coordinates with respect to their basis vectors as:

$$X = X_0 x_0 + X_1 x_1 + X_2 x_2 + X_3 x_3 \tag{4.1}$$

The coordinates are given by:

$$\begin{cases} x_0 = -\,(X, X_0) \\ x_1 = \quad (X, X_1) \\ x_2 = \quad (X, X_2) \\ x_3 = \quad (X, X_3) \end{cases} \tag{4.2}$$

where (a, b) represents the scalar product of vectors a and b (see Sect. 2.3.3).

Note that the temporal basis vector X_0 does not behave the same as the spatial basis vectors. The scalar product (X_0, X_0) produces the value *negative* one instead of, as for the other basis vectors, a value of positive one.

The basis vectors of Minkowskian geometry can themselves be represented with Clifford units as:

$$\begin{cases} X_0 = i e_0 \\ X_1 = e_1 \\ X_2 = e_2 \\ X_3 = e_3 \end{cases} \tag{4.3}$$

The temporal basis vector incorporates the imaginary unit i to ensure that the scalar product $(X_0, X_0) = \overline{X}_0 X_0 = \overline{i e_0}\, i e_0 = i\, \overline{e_0}\, i e_0 = i(-e_0)\, i e_0$, where $\overline{a}$ is the Clifford (not complex) conjugate of a (see Sect. 2.3.1), produces the negative value as required.

4.2 Transformation Between Frames

Suppose now that the Clifford–Minkowskian frame $\mathbb{X} = \{X_0, X_1, X_2, X_3\} = \{i e_0, e_1, e_2, e_3\}$ is fixed in space at any chosen point, and a second Minkowskian frame $\mathbb{Y} = \{Y_0, Y_1, Y_2, Y_3\}$ is passing through the same point at a constant velocity with respect to $\mathbb{X}$. In this case the two frames and their corresponding sets of coordinates x_0, x_1, x_2, x_3 and y_0, y_1, y_2, y_3 are related by the Lorentz transformation. The forward transform $w = L u \widetilde{L}$ is used when mapping from the coordinates in the $\mathbb{X}$ frame to the coordinates in the $\mathbb{Y}$ frame. The inverse transform $u = \widetilde{L} w L$ is used when mapping from the basis vectors in the $\mathbb{X}$ frame to the basis vectors in the $\mathbb{Y}$ frame.

The Lorentz transformation and its inverse can be interpreted geometrically as hyperbolic rotations, *i.e.* rotations through (opposite) angles of imaginary value. If the motion of the $\mathbb{Y}$ frame is in the direction of the X_1 basis vector then these hyperbolic rotations dictate mappings between the X_0 and X_1 basis vectors of the $\mathbb{X}$ frame and the Y_0 and Y_1 basis vectors of the $\mathbb{Y}$ frame.

The rotation $\Theta(u) = Lu\widetilde{L}$ in the X_0X_1 plane is specified (as in Sect. 3.3) by Clifford unit vectors, here $n_\psi = n(\psi) = X_0 \cosh(\psi/2) + X_1 \sinh(\psi/2)$ for both $\psi = 0$ and $\psi = a$:

$$\begin{cases} n_0 = ie_0 & = X_0 \\ n_a = ie_0 \cosh(a/2) + e_1 \sinh(a/2) & = X_0 \cosh(a/2) + X_1 \sinh(a/2) \end{cases} \tag{4.4}$$

The amount of rotation is governed by the arbitrary constant a, which sets the magnitude of the hyperbolic angle between the two vectors n_0, n_a. The rotation itself covers twice that angle, in a direction governed by the order in which the vectors are applied. In the usual Clifford fashion, any vector u is rotated by a reflection $R_1(u)$ of all components in the direction of one of the vectors (say n_a):

$$v = R_1(u) = n_a u n_a \tag{4.5}$$

followed by a reflection $R_2(v)$ of all components in the direction of the other vector (in this case n_0):

$$w = \Theta(u) = R_2(v) = n_0 v n_0 = n_0 n_a u n_a n_0 = Lu\widetilde{L} \tag{4.6}$$

The transformation can be reversed by (re)applying the reflections in reverse order:

$$\Theta^{-1}(\Theta(u)) = \widetilde{L}wL = n_a n_0 (n_0 n_a u n_a n_0) n_0 n_a = \widetilde{L}(Lu\widetilde{L})L = u \tag{4.7}$$

The final equality is readily seen by noticing from Table 2.2 that $\widetilde{L}L = L\widetilde{L} = 1$ when L is a unit bivector, or by noticing that the square of a unit vector n is $n^2 = -1$ from Eq. 1.17, or by noticing that the reflection of any object twice in the same direction (such as $n_0 n_0 v n_0 n_0$) recovers the original object.

4.2.1 Related Frames

Applying the inverse Lorentz transformation to the basis of the $\mathbb{X}$ frame gives the basis of the moving frame $\mathbb{Y}$ as:

$$\begin{cases} Y_0 = \widetilde{L}X_0L = ie_0 \cosh a + e_1 \sinh a = X_0 \cosh a + X_1 \sinh a \\ Y_1 = \widetilde{L}X_1L = ie_0 \sinh a + e_1 \cosh a = X_0 \sinh a + X_1 \cosh a \\ Y_2 = \widetilde{L}X_2L = e_2 \qquad\qquad\qquad\qquad = X_2 \\ Y_3 = \widetilde{L}X_3L = e_3 \qquad\qquad\qquad\qquad = X_3 \end{cases} \tag{4.8}$$

In the $\mathbb{Y}$ frame the location of the spatial origin O_Y, with spatial coordinates $y_1 = y_2 = y_3 = 0$, is given for any value y_0 of the time coordinate by the space–time vector:

$$\begin{aligned} O_Y &= Y_0 y_0 + Y_1 0 + Y_2 0 + Y_3 0 = Y_0 y_0 \\ &= (i e_0 \cosh a + e_1 \sinh a) y_0 = (X_0 \cosh a + X_1 \sinh a) y_0 \end{aligned} \tag{4.9}$$

The coordinates of the same location in space–time is seen as a different set of coordinates in the $\mathbb{X}$ frame. The $\mathbb{X}$ frame coordinates are obtained by using the scalar product to map the space–time vector O_Y onto the $\mathbb{X}$ frame:

$$\begin{cases} x_0 = -(O_Y, X_0) = (\cosh a) y_0 \\ x_1 = (O_Y, X_1) = (\sinh a) y_0 \\ x_2 = (O_Y, X_2) = 0 \\ x_3 = (O_Y, X_3) = 0 \end{cases} \tag{4.10}$$

The origin O_Y is moving in both space (along the X_1 axis) and in time (along the X_0 axis) at a constant velocity V_X seen from the $\mathbb{X}$ frame as:

$$V_X = \frac{dx_1}{dx_0} = \frac{x_1}{x_0} = \tanh a \tag{4.11}$$

From the other viewpoint, the spatial origin O_X of the $\mathbb{X}$ frame:

$$O_X = X_0 x_0 + X_1 0 + X_2 0 + X_3 0 = X_0 x_0 = i e_0 x_0 \tag{4.12}$$

with $\mathbb{X}$ frame coordinates $x_1 = x_2 = x_3 = 0$, has coordinates as seen from the $\mathbb{Y}$ frame:

$$\begin{cases} y_0 = -(O_X, Y_0) = (\cosh a) x_0 \\ y_1 = (O_X, Y_1) = -(\sinh a) x_0 \\ y_2 = (O_X, Y_2) = 0 \\ y_3 = (O_X, Y_3) = 0 \end{cases} \tag{4.13}$$

The origin O_X is moving in both space (along the Y_1 axis) and in time (along the Y_0 axis) at a constant velocity V_Y seen from the $\mathbb{Y}$ frame as:

$$V_Y = \frac{dy_1}{dy_0} = \frac{y_1}{y_0} = -\tanh a \tag{4.14}$$

4.2.2 Related Coordinates

Any point P can be equally well expressed in terms of the coordinates x_0, x_1, x_2, x_3 of the $\mathbb{X}$ frame or the coordinates y_0, y_1, y_2, y_3 in the $\mathbb{Y}$ frame:

$$P = \begin{cases} X = X_0x_0 + X_1x_1 + X_2x_2 + X_3x_3 \\ Y = Y_0y_0 + Y_1y_1 + Y_2y_2 + Y_3y_3 \end{cases} \tag{4.15}$$

If the point $P = X$ is known in terms of its coordinates in the $\mathbb{X}$ frame, the forward transform then produces the coordinates in the $\mathbb{Y}$ frame:

$$\begin{aligned} LX\widetilde{L} &= L(Y_0y_0 + Y_1y_1 + Y_2y_2 + Y_3y_3)\widetilde{L} \\ &= LY_0\widetilde{L}y_0 + LY_1\widetilde{L}y_1 + LY_2\widetilde{L}y_2 + LY_3\widetilde{L}y_3 = X_0y_0 + X_1y_1 + X_2y_2 - X_3y_3 \end{aligned} \tag{4.16}$$

so that:

$$\begin{cases} y_0 = -\,(LX\widetilde{L}, X_0) \\ y_1 = (LX\widetilde{L}, X_1) \\ y_2 = (LX\widetilde{L}, X_2) \\ y_3 = (LX\widetilde{L}, X_3) \end{cases} \tag{4.17}$$

4.3 Derivatives in Space–Time

Figure 4.1 shows the temporal basis vectors X_0, Y_0 and spatial basis vectors X_1, Y_1 for two Minkowskian frames $\mathbb{X} = \{X_0, X_1, X_2, X_3\}$ and $\mathbb{Y} = \{Y_0, Y_1, Y_2, Y_3\}$. The values in parentheses are the coordinates of certain points along each of the four basis vectors, in terms of the corresponding frame. The remaining spatial basis vectors and corresponding coordinates are not shown since they play no consequential role in the discussion.

The $\mathbb{X}$ frame is nominally fixed, and the spatial origin of the $\mathbb{Y}$ frame is moving in the positive X_1 direction with a velocity $V_X = \tanh a$. The basis vectors in the $\mathbb{Y}$ frame are then (as in Eq. 4.8):

$$\begin{pmatrix} Y_0 & Y_1 \end{pmatrix} = \begin{pmatrix} X_0 & X_1 \end{pmatrix} \begin{pmatrix} \cosh a & \sinh a \\ \sinh a & \cosh a \end{pmatrix} \tag{4.18}$$

In terms of the $\mathbb{Y}$ frame the basis vectors of the $\mathbb{X}$ frame are by matrix inversion:

$$\begin{pmatrix} X_0 & X_1 \end{pmatrix} = \begin{pmatrix} Y_0 & Y_1 \end{pmatrix} \begin{pmatrix} \cosh a & -\sinh a \\ -\sinh a & \cosh a \end{pmatrix} \tag{4.19}$$

Any point P can equally well be expressed in terms of the coordinates x_0, x_1 of the $\mathbb{X}$ frame, or the coordinates y_0, y_1 of the $\mathbb{Y}$ frame:

$$\begin{aligned} P = \quad & X_0x_0 + X_1x_1 = Y_0y_0 + Y_1y_1 \\ & \begin{pmatrix} X_0 & X_1 \end{pmatrix} \begin{pmatrix} x_0 \\ x_1 \end{pmatrix} = \begin{pmatrix} Y_0 & Y_1 \end{pmatrix} \begin{pmatrix} y_0 \\ y_1 \end{pmatrix} \end{aligned} \tag{4.20}$$

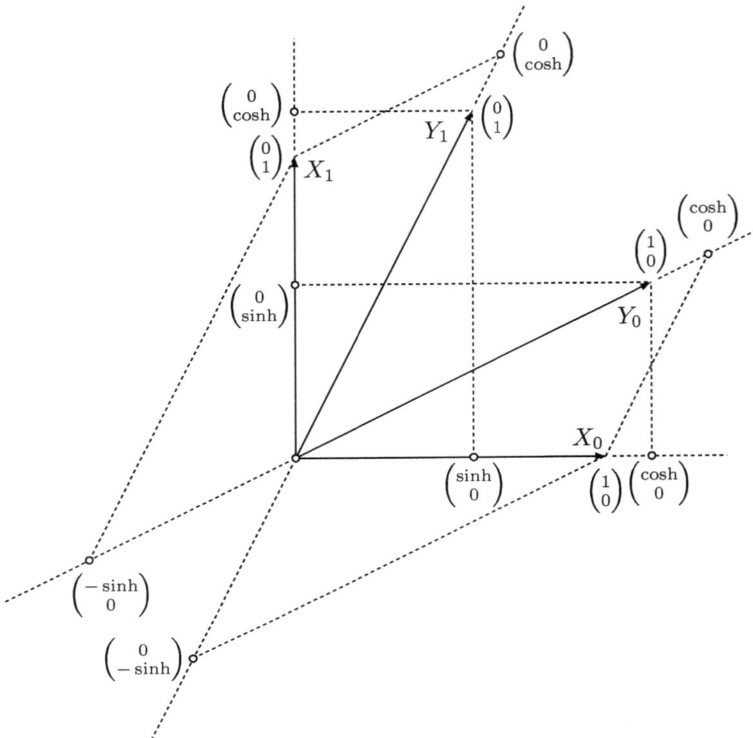

Fig. 4.1 Lorentz transform between frame $\mathbb{X} = \{X_0, X_1\}$ and frame $\mathbb{Y} = \{Y_0, Y_1\}$

Substituting for X_0, X_1 from Eq. 4.19 gives the transformation from coordinates in the $\mathbb{X}$ frame to the $\mathbb{Y}$ frame:

$$\begin{pmatrix} y_0 \\ y_1 \end{pmatrix} = \begin{pmatrix} \cosh a & -\sinh a \\ -\sinh a & \cosh a \end{pmatrix} \begin{pmatrix} x_0 \\ x_1 \end{pmatrix} \tag{4.21}$$

As the point P moves, the coordinates in both the $\mathbb{X}$ and $\mathbb{Y}$ frames change accordingly. Differential changes in coordinates of the $\mathbb{Y}$ frame with respect to the coordinates of the $\mathbb{X}$ frame are given by:

$$\begin{pmatrix} \frac{\partial y_0}{\partial x_0} & \frac{\partial y_0}{\partial x_1} \\ \frac{\partial y_1}{\partial x_0} & \frac{\partial y_1}{\partial x_1} \end{pmatrix} = \begin{pmatrix} \cosh a & -\sinh a \\ -\sinh a & \cosh a \end{pmatrix} \tag{4.22}$$

When a function F is defined in the vicinity of the point P, its partial derivatives can be calculated with respect to the coordinates in either the $\mathbb{X}$ frame or the $\mathbb{Y}$ frame, and then converted to the other frame by the chain rule for partial differentiation.

Utilising the partial derivatives for the Lorentz transformation as given by Eq. 4.22 then leads to:

$$\begin{pmatrix} \frac{\partial F}{\partial x_0} \\ \frac{\partial F}{\partial x_1} \end{pmatrix} = \begin{pmatrix} \frac{\partial y_0}{\partial x_0} & \frac{\partial y_1}{\partial x_0} \\ \frac{\partial y_0}{\partial x_1} & \frac{\partial y_1}{\partial x_1} \end{pmatrix} \begin{pmatrix} \frac{\partial F}{\partial y_0} \\ \frac{\partial F}{\partial y_1} \end{pmatrix} = \begin{pmatrix} \cosh a & -\sinh a \\ -\sinh a & \cosh a \end{pmatrix}^T \begin{pmatrix} \frac{\partial F}{\partial y_0} \\ \frac{\partial F}{\partial y_1} \end{pmatrix} \tag{4.23}$$

Rather than dealing always with partial differential operators it is more convenient to construct from them a complete differential operator $\mathcal{D}$ which works equally well and consistently in both the $\mathbb{X}$ frame and the $\mathbb{Y}$ frame. To work consistently the result from the $\mathbb{X}$ frame must agree when converted into the $\mathbb{Y}$ frame and *vice versa*. Specifically:

$$\mathcal{D}F = X_0 a_0 \frac{\partial F}{\partial x_0} + X_1 a_1 \frac{\partial F}{\partial x_1} = Y_0 a_0 \frac{\partial F}{\partial y_0} + Y_1 a_1 \frac{\partial F}{\partial y_1} \tag{4.24}$$

with the unknown constants a_0 and a_1 chosen to establish the equality:

$$\begin{aligned} (X_0\ X_1) \begin{pmatrix} a_0 & 0 \\ 0 & a_1 \end{pmatrix} \begin{pmatrix} \frac{\partial F}{\partial x_0} \\ \frac{\partial F}{\partial x_1} \end{pmatrix} &= (Y_0\ Y_1) \begin{pmatrix} a_0 & 0 \\ 0 & a_1 \end{pmatrix} \begin{pmatrix} \frac{\partial F}{\partial y_0} \\ \frac{\partial F}{\partial y_1} \end{pmatrix} \\ (X_0\ X_1) \begin{pmatrix} a_0 & 0 \\ 0 & a_1 \end{pmatrix} \begin{pmatrix} \cosh a & -\sinh a \\ -\sinh a & \cosh a \end{pmatrix} \begin{pmatrix} \frac{\partial F}{\partial y_0} \\ \frac{\partial F}{\partial y_1} \end{pmatrix} &= (X_0\ X_1) \begin{pmatrix} \cosh a & \sinh a \\ \sinh a & \cosh a \end{pmatrix} \begin{pmatrix} a_0 & 0 \\ 0 & a_1 \end{pmatrix} \begin{pmatrix} \frac{\partial F}{\partial y_0} \\ \frac{\partial F}{\partial y_1} \end{pmatrix} \\ \begin{pmatrix} a_0 \cosh a & -a_0 \sinh a \\ -a_1 \sinh a & a_1 \cosh a \end{pmatrix} &= \begin{pmatrix} (\cosh a)a_0 & (\sinh a)a_1 \\ (\sinh a)a_0 & (\cosh a)a_1 \end{pmatrix} \\ a_0 &= -a_1 \end{aligned} \tag{4.25}$$

The equality in the last line must be satisfied, except in the trivial case of zero relative velocity V_X, for which $a = \tanh^{-1} V_X = 0$. Here a value $a_1 = 1$ is chosen for simplicity, in which case $a_0 = -1$. The differential operator is then given by:

$$\mathcal{D} = -X_0 \frac{\partial}{\partial x_0} + X_1 \frac{\partial}{\partial x_1} \tag{4.26}$$

Reintroducing the two dimensions left aside from the discussion until now, gives the full form of the differential operator in the case of one time and three spatial dimensions:

$$\mathcal{D} = -X_0 \frac{\partial}{\partial x_0} + X_1 \frac{\partial}{\partial x_1} + X_2 \frac{\partial}{\partial x_2} + X_3 \frac{\partial}{\partial x_3} \tag{4.27}$$

This differential operator plays the role of the gradient in Minkowskian space–time. This form of differential operator is the one which must be used when casting the temporal version of Maxwell's equations in the form of Clifford algebra (see Sect. 5.1.2). The negative sign is mandatory, not optional. Failing to employ the negative sign violates Einstein's theory of special relativity.

4.4 Physical Units

The Minkowskian geometry of space–time described in Sects. 4.1 and 4.3 treats all of the unit vectors in the frame $\mathbb{X}$ as if they are of the same kind. Specifically, as vectors in a geometric space, they all have norms which represent some kind of distances. More particularly, they all take the same kind of physical units.

In engineering it is customary to adopt different units: units of length (such as metres) for the spatial basis vectors and units of time (such as seconds) for the temporal basis vector. Consider therefore replacing the time-like coordinate x_0 with a rescaled variable t_x:

$$x_0 = c\, t_x \tag{4.28}$$

for some scale factor c. It is now possible to adopt physical units of length for coordinates x_0, x_1, x_2, x_3, and a unit of time for the variable t_x, provided the scale factor c has units of length/time = velocity.

In terms of the new (scaled) temporal variable t_x, the velocity of the origin O_Y of the moving frame $\mathbb{Y}$ (see Eq. 4.11) is:

$$v_x = \frac{dx_1}{dt_x} = \frac{dx_1}{dx_0}\frac{dx_0}{dt_x} = c\ \tanh a = c\, V_X \tag{4.29}$$

with units of length/time. Since the hyperbolic tangent function takes no values outside the range ± 1, it is never possible from the $\mathbb{X}$ frame to observe another frame $\mathbb{Y}$ moving at any velocity greater than c. Exactly the same is true of the $\mathbb{X}$ frame when viewed from the $\mathbb{Y}$ frame.

The new time variable t_x must also be incorporated into the space–time differential operator $\mathcal{D}$ from Eq. 4.27:

$$\begin{aligned}
\mathcal{D} &= -X_0\tfrac{\partial}{\partial t_x}\left(\tfrac{dt_x}{dx_0}\right) + X_1\tfrac{\partial}{\partial x_1} + X_2\tfrac{\partial}{\partial x_2} + X_3\tfrac{\partial}{\partial x_3} \\
&= -\tfrac{1}{c}X_0\tfrac{\partial}{\partial t_x} + X_1\tfrac{\partial}{\partial x_1} + X_2\tfrac{\partial}{\partial x_2} + X_3\tfrac{\partial}{\partial x_3} \\
&= -\tfrac{i}{c}e_0\tfrac{\partial}{\partial t} + e_1\tfrac{\partial}{\partial x} + e_2\tfrac{\partial}{\partial y} + e_3\tfrac{\partial}{\partial z}
\end{aligned} \tag{4.30}$$

where $t = t_x$, $x = x_1$, $y = x_2$ and $z = x_3$.

4.5 Electromagnetic Field

The differential operator $\mathcal{D}$ in Eq. 4.27 or 4.30 accommodates Einstein's theory of special relativity for any function F. In the particular case where $F = \mathcal{F}$ is the electromagnetic field, the form of that field $\mathcal{F}$ is fixed by the additional requirement to reproduce Maxwell's equations under the influence of the operator $\mathcal{D}$. This is achieved (see Sect. 5.1.2) when the electromagnetic field $\mathcal{F}$ is written in the $\mathbb{X}$ frame as:

$$\mathcal{F} = \sqrt{\mu} H\sigma - i\sqrt{\epsilon} E e_0 = -X_1 X_2 X_3 \sqrt{\mu} H + X_0 \sqrt{\epsilon} E \tag{4.31}$$

where $H = X_1 H_x + X_2 H_y + X_3 H_z = H_x e_1 + H_y e_2 + H_z e_3$ is the magnetic field in the $\mathbb{X}$ frame and $E = X_1 E_x + X_2 E_y + X_3 E_z = E_x e_1 + E_y e_2 + E_z e_3$ is the electric field in the $\mathbb{X}$ frame.

The same field viewed from the moving $\mathbb{Y}$ frame appears to have different values of electric and magnetic fields. Electric fields, which arise naturally from fixed charges in the $\mathbb{X}$ frame, exhibit an aspect of magnetic behaviour in the $\mathbb{Y}$ frame, since in that frame the charges appear to be moving. Similarly, magnetic fields, which arise from moving charges in the $\mathbb{X}$ frame, exhibit an aspect of electric character in the $\mathbb{Y}$ frame.

The Lorentz transformation can be used to calculate the components of the field in the moving $\mathbb{Y}$ frame. The method is the same as used in Sect. 4.2.2 for coordinates. It is only necessary to note that any products of basis vectors in the $\mathbb{Y}$ frame map into the corresponding products of basis vectors for the $\mathbb{X}$ frame, as in the case $LY_0Y_1\widetilde{L} = LY_0\widetilde{L}LY_1\widetilde{L} = X_0X_1$.

It is therefore possible to extract the components of the electric and magnetic fields E', H' in the $\mathbb{Y}$ frame using the Lorentz transformation, the basis vectors associated with each component in the $\mathbb{X}$ frame and the scalar product as follows:

$$\begin{cases} \sqrt{\mu}H'_x = -(L\mathcal{F}\widetilde{L}, X_1X_2X_3X_1) = \sqrt{\mu}H_x \\ \sqrt{\mu}H'_y = -(L\mathcal{F}\widetilde{L}, X_1X_2X_3X_2) = \sqrt{\mu}H_y \cosh a + \sqrt{\epsilon}E_z \sinh a \\ \sqrt{\mu}H'_z = -(L\mathcal{F}\widetilde{L}, X_1X_2X_3X_3) = \sqrt{\mu}H_z \cosh a - \sqrt{\epsilon}E_y \sinh a \\ \sqrt{\epsilon}E'_x = -(L\mathcal{F}\widetilde{L}, X_0X_1) \quad = \sqrt{\epsilon}E_x \\ \sqrt{\epsilon}E'_y = -(L\mathcal{F}\widetilde{L}, X_0X_2) \quad = \sqrt{\epsilon}E_y \cosh a - \sqrt{\mu}H_z \sinh a \\ \sqrt{\epsilon}E'_z = -(L\mathcal{F}\widetilde{L}, X_0X_3) \quad = \sqrt{\epsilon}E_z \cosh a + \sqrt{\mu}H_y \sinh a \end{cases} \tag{4.32}$$

The negative signs in the central column of Eq. 4.32 come from different places. In the equations for H' they come from the minus sign in Eq. 4.31. In the equations for E' they appear in accordance with the negative scalar product of the basis vector X_0, as in Eq. 4.2.

4.6 Exercises

Q1. With $X = 3X_0 + 4X_1$ verify that the scalar product $-(X, X_0)$ in Eq. 4.2 recovers the value of the coordinate x_0.

Q2. With $X = 3X_0 + 4X_1$ verify that the scalar product (X, X_1) in Eq. 4.2 recovers the value of the coordinate x_1.

Q3. With $X = 3X_0 + 4X_1$ calculate the value of the norm $\|X\| = (X, X)$.

Q4. With $X = 3X_1 + 4X_2$ calculate the value of the norm $\|X\| = (X, X)$.

Q5. Show that $\overline{ab} = \widetilde{ab} = ba$ for any two vectors a, b.

Q6. Check the electric and magnetic field Lorentz mappings in Eq. 4.32 against any standard reference.

Chapter 5
Classical Electromagnetism

Electromagnetism is a physical phenomenon caused by electric charges, both static and moving. Maxwell's equations are a mathematical model based on experimental observations of electromagnetic phenomena in the 19th century.

Maxwell's equations can be written as a set of four first order partial differential equations with two coefficients μ and ϵ (as in Eq. 5.1). These coefficients are governed by the material properties of the regions in which the equations are observed to apply.

Although materials are composed of atoms they are typically treated for the purposes of electromagnetism at a macroscopic level as being composed of some continuous substance. The corresponding continuous magnetic and electric material properties μ and ϵ can be interpreted as convenient kinds of statistical and quantum mechanical averages which mathematically represent at the macroscopic level the underlying atomic construction of the material.

In those physical situations for which they are suited, Maxwell's equations predict very accurately many of the physical manifestations of electromagnetic behaviour. Collectively, manifestations in this category are referred to under the name of *Classical Electromagnetism*. Other physical manifestations related to particular processes at the atomic level or under unusual physical conditions are more accurately predicted by the more detailed equations of *Quantum Electrodynamics*. Examples of behaviour outside the realm of classical electromagnetism include the quantum nature of the photoelectric effect, the discrete spectral nature of radiation emitted from heated material, materials in the form of plasma or *not* in a state of thermal equilibrium, and situations involving general relativistic effects such as strong gravitation (black holes).

Most of electrical engineering lies within the scope of classical electromagnetism. In engineering practice it is not a serious limitation to restrict attention further to cases where the coefficients μ and ϵ are for the most part constant. Maxwell's equations then apply only to regions where the material properties take values which are uniform, or where regions meet at interfaces at which the material properties jump abruptly from one uniform value to another. When subject to these restrictions Maxwell's

A. Seagar, *Application of Geometric Algebra to Electromagnetic Scattering*,
DOI 10.1007/978-981-10-0089-8_5

equations are not suited for regions where the material properties vary continuously with position.

The methods described herein are applied to Maxwell's equations with constant coefficients, and relate to the phenomena encompassed by classical (non-quantum) electromagnetism.

5.1 Time Domain

5.1.1 Vector Calculus

In the notation of vector calculus Maxwell's equations are written [14, 29] as a system of four first order partial differential equations:

$$\begin{cases} \nabla \cdot (\epsilon \mathbf{E}) = \rho \\ \nabla \times \mathbf{H} - \frac{\partial(\epsilon \mathbf{E})}{\partial t} = \mathbf{J} \\ \nabla \times \mathbf{E} + \frac{\partial(\mu \mathbf{H})}{\partial t} = 0 \\ \nabla \cdot (\mu \mathbf{H}) = 0 \end{cases} \tag{5.1}$$

where $\mathbf{E}$ and $\mathbf{H}$ are the electric and magnetic fields respectively, ρ and $\mathbf{J}$ are the charge and current source densities respectively, and ϵ and μ are the electric permittivity and magnetic permeability respectively. The divergence $\nabla\cdot$ and the curl $\nabla\times$ are partial differential operators with respect to space which produce the average radial and tangential variations of the fields on which they operate.

At interfaces where the material properties ϵ and μ jump abruptly from one uniform value to the other, provided that sources are absent, the normal components of the electric and magnetic flux densities $\mathbf{D} = \epsilon\mathbf{E}$ and $\mathbf{B} = \mu\mathbf{H}$ and the tangential components of the fields $\mathbf{E}$ and $\mathbf{H}$ are continuous. In the case that sources are present at the interface, there is a step change in the value of flux density or field proportional to the strength of the source.

5.1.2 Clifford Algebra

Casting Maxwell's equations into the framework of Clifford algebra involves firstly creating Clifford entities as in Table 5.1 equivalent to those used in vector calculus. The symbols for the new variables are chosen sufficiently close to the form of the originals so that they can be readily identified. The scalar forms are identical.

Secondly, the combined temporal and spatial differential operator $\mathcal{D}$ is constructed in the form:

$$\mathcal{D} = -\tfrac{i}{c} e_0 \tfrac{\partial}{\partial t} + e_1 \tfrac{\partial}{\partial x} + e_2 \tfrac{\partial}{\partial y} + e_3 \tfrac{\partial}{\partial z} = \nabla - \tfrac{i}{c} e_0 \tfrac{\partial}{\partial t} \tag{5.2}$$

Table 5.1 Notation for electromagnetic quantities and operators

Type	Entity	Vector calculus	Clifford algebra
Scalar	Magnetic permeability	μ	μ
	Electric permittivity	ϵ	ϵ
	Charge density	ρ	ρ
	Electric potential	ϕ	ϕ
	Temporal derivative	$\partial/\partial t$	$\partial/\partial t$
Vector	Magnetic field	$\mathbf{H}$	H
	Electric field	$\mathbf{E}$	E
	Current density	$\mathbf{J}$	J
	Magnetic potential	$\mathbf{A}$	A
	Divergence	$\nabla\cdot$	$\boldsymbol{\nabla}\cdot$
	Curl	$\nabla\times$	$\boldsymbol{\nabla}\times$

so that, as in Chap. 4, it adheres to Einstein's theory of special relativity. Thirdly, the electromagnetic field $\mathcal{F}$ is constructed in the form:

$$\mathcal{F} = \sqrt{\mu}H\sigma - i\sqrt{\epsilon}Ee_0 \tag{5.3}$$

so that the Clifford product $\mathcal{DF}$, when split into four independent quaternion components, matches to within multiplicative constants the left hand side of Eq. 5.1. Finally, the source field $\mathcal{S}$ is constructed in the form:

$$\mathcal{S} = \sqrt{\mu}J + \frac{i}{\sqrt{\epsilon}}\rho e_0 \tag{5.4}$$

so that its Clifford value, when split into quaternion components, matches to within the same multiplicative constants as the field, the right hand side of Eq. 5.1.

The final result for Maxwell's equations in the framework of Clifford algebra is the single first order ordinary differential equation:

$$\mathcal{DF} = \mathcal{S} \tag{5.5}$$

Consistency of the result with the Eq. 5.1 in vector form is a matter of expanding Eq. 5.5 using Clifford algebra and separating the independent components. Firstly expanding Clifford products in the cross-dot form of Eq. 1.16 gives:

$$\begin{aligned}
\mathcal{DF} &= \mathcal{S} \\
(\boldsymbol{\nabla} - \tfrac{i}{c}e_0\tfrac{\partial}{\partial t})(\sqrt{\mu}H\sigma - i\sqrt{\epsilon}Ee_0) &= \\
\sqrt{\mu}(-\boldsymbol{\nabla}\cdot H + \boldsymbol{\nabla}\times H\sigma)\sigma - i\sqrt{\epsilon}(-\boldsymbol{\nabla}\cdot E + \boldsymbol{\nabla}\times E\sigma)e_0 - \tfrac{i}{c}e_0\sqrt{\mu}\tfrac{\partial H}{\partial t}\sigma - \tfrac{1}{c}e_0\sqrt{\epsilon}\tfrac{\partial E}{\partial t}e_0 &= \\
-\sqrt{\mu}\boldsymbol{\nabla}\cdot H\sigma + \sqrt{\mu}\boldsymbol{\nabla}\times H + i\sqrt{\epsilon}\boldsymbol{\nabla}\cdot Ee_0 + i\sqrt{\epsilon}\boldsymbol{\nabla}\times Ee_0\sigma + i\sqrt{\epsilon}\mu\tfrac{\partial H}{\partial t}e_0\sigma - \sqrt{\mu}\epsilon\tfrac{\partial E}{\partial t} &= \sqrt{\mu}J + \tfrac{i}{\sqrt{\epsilon}}\rho e_0
\end{aligned} \tag{5.6}$$

The equation then separates according to the quaternion units ($\mathbf{1}$, $\mathbf{I}_2 = e_0$, $\mathbf{J}_2 = -i\sigma$, $\mathbf{K}_2 = -ie_0\sigma$) as a result of their linear independence, giving:

$$\left\{\begin{array}{lcl} \frac{i}{\sqrt{\epsilon}}(\epsilon\nabla \cdot E)e_0 & = & \frac{i}{\sqrt{\epsilon}}(\rho)\, e_0 \\ \sqrt{\mu}(\nabla \times H - \epsilon\frac{\partial E}{\partial t}) & = & \sqrt{\mu}(J) \\ -\sqrt{\epsilon}(\nabla \times E + \mu\frac{\partial H}{\partial t})(-ie_0\sigma) & = & 0 \\ -\frac{i}{\sqrt{\mu}}(\mu\nabla \cdot H)(-i\sigma) & = & 0 \end{array}\right. \tag{5.7}$$

Reference to Eq. 5.1 verifies for constant values of μ and ϵ that each of these equations is a copy of one of Maxwell's equations in its vector form, scaled outside the parentheses on the left by a constant and on the right by a quaternion unit.

Applying the same differential operator $\mathcal{D}$ to a Clifford potential function:

$$\mathcal{P} = \frac{1}{\sqrt{\mu}}A + i\sqrt{\epsilon}\phi e_0 \tag{5.8}$$

containing the vector magnetic potential A and the scalar electric potential ϕ reproduces the electromagnetic field $\mathcal{F}$:

$$\begin{gathered} \mathcal{D}\mathcal{P} = \mathcal{F} \\ (\nabla - \tfrac{i}{c}e_0\tfrac{\partial}{\partial t})(\tfrac{1}{\sqrt{\mu}}A + i\sqrt{\epsilon}\phi e_0) = \\ \tfrac{1}{\sqrt{\mu}}(-\nabla \cdot A + \nabla \times A\sigma) + i\sqrt{\epsilon}(\nabla\phi)e_0 - \tfrac{i}{c}e_0\tfrac{1}{\sqrt{\mu}}\tfrac{\partial A}{\partial t} + \tfrac{1}{c}e_0\sqrt{\epsilon}\tfrac{\partial \phi}{\partial t}e_0 = \\ -\tfrac{1}{\sqrt{\mu}}\nabla \cdot A + \tfrac{1}{\sqrt{\mu}}\nabla \times A\sigma + i\sqrt{\epsilon}\nabla\phi e_0 + i\sqrt{\epsilon}\tfrac{\partial A}{\partial t}e_0 - \sqrt{\mu}\epsilon\tfrac{\partial \phi}{\partial t} = \sqrt{\mu}H\sigma - i\sqrt{\epsilon}Ee_0 \end{gathered} \tag{5.9}$$

The equation separates according to the quaternion units ($\mathbf{1}$, $\mathbf{I}_2 = e_0$, $\mathbf{J}_2 = -i\sigma$, $\mathbf{K}_2 = -ie_0\sigma$) as a result of their linear independence, giving:

$$\left\{\begin{array}{lcl} -i\sqrt{\epsilon}(-\nabla\phi - \frac{\partial A}{\partial t})e_0 & = & -i\sqrt{\epsilon}(E)e_0 \\ -\frac{1}{\sqrt{\mu}}(\nabla \cdot A + \frac{1}{c^2}\frac{\partial \phi}{\partial t}) & = & 0 \\ \frac{i}{\sqrt{\mu}}(\nabla \times A)(-i\sigma) & = & \frac{i}{\sqrt{\mu}}(\mu H)\,(-i\sigma) \end{array}\right. \tag{5.10}$$

The central equation is the Lorentz condition (*cf.* [29]), which determines the value of the divergence of the magnetic vector potential. It appears here by necessity rather than by choice, a result of already having required that the differential operator $\mathcal{D}$ adhere to special relativity in the form of the Lorentz transformation (see Chap. 4).

Reference to a standard text on electromagnetism [29] verifies that the other two equations are copies of the potential equations in their vector form, scaled on the left by a constant and on the right by a quaternion unit.

Table 5.2 Electromagnetic quantities in the form of Clifford variables, and quaternions ($\mathfrak{A}$, $\mathfrak{B}$, $\mathfrak{C}$, $\mathfrak{D}$) split into their scalar (a, b, c, d) and vector (A, B, C, D) components

	Clifford	Quaternion							
		$\mathfrak{A}$		$\mathfrak{B}$		$\mathfrak{C}$		$\mathfrak{D}$	
		a	A	b	B	c	C	d	D
Operator	$\mathcal{D} = \nabla - \frac{i}{c} e_0 \frac{\partial}{\partial t}$	0	0	$-\frac{i}{c}\frac{\partial}{\partial t}$	0	0	$i\nabla$	0	0
Field	$\mathcal{F} = \sqrt{\mu} H\sigma - i\sqrt{\epsilon} E e_0$	0	$\sqrt{\mu} H$	0	0	0	0	0	$-\sqrt{\epsilon} E$
Source	$\mathcal{S} = \sqrt{\mu} J + \frac{i}{\sqrt{\epsilon}} \rho e_0$	0	0	$\frac{i}{\sqrt{\epsilon}} \rho$	0	0	$i\sqrt{\mu} J$	0	0
Potential	$\mathcal{P} = \frac{1}{\sqrt{\mu}} A + i\sqrt{\epsilon}\phi e_0$	0	0	$i\sqrt{\epsilon}\phi$	0	0	$\frac{i}{\sqrt{\mu}} A$	0	0

Differentiating Eq. 5.9 gives:

$$\mathcal{D}^2 \mathcal{P} = \mathcal{S} \tag{5.11}$$

which is recognised as the wave equation for the electromagnetic potential in Clifford form after noting that:

$$\mathcal{D}^2 = -(\nabla^2 - \frac{1}{c^2}\frac{\partial^2}{\partial t^2}) \tag{5.12}$$

is the negative of the d'Alembertian operator.

Table 5.2 summarises the different electromagnetic quantities expressed (see Sect. 1.6) as Clifford numbers on the left and as quaternions on the right. When expressed in terms of Clifford algebra each quantity requires only one number. In comparison each quantity requires two independent quaternions, with four independent ones required to accommodate all of the quantities.

5.2 Splitting the Field

Splitting of single Clifford-valued electromagnetic field equations such as Eqs. 5.6 and 5.9, into four components according to quaternion units, as in Eqs. 5.7 and 5.10, is handled formally in the framework of Clifford algebra by introducing operators specifically designed for that purpose. Here, two pairs of splitting operators are required, to successively split the field equations first in half and then in half again. In other sections, other splitting operators are used for other purposes.

Splitting operators can be constructed as projection operators from reflection operators using the same method as in Sect. 3.1, and as demonstrated in Sects. 3.2.1 and 3.2.3. They are used there to split the field there into normal and tangential components.

Here, the first step is to define two reflection operators:

$$Mu = \sigma u \sigma \tag{5.13}$$
$$Nu = e_0 u e_0 \tag{5.14}$$

From those, two pairs of projection operators are constructed:

$$\begin{cases} S = \frac{1}{2}(I + M) & \text{space} \\ T = \frac{1}{2}(I - M) & \text{time} \end{cases} \tag{5.15}$$

and:

$$\begin{cases} V = \frac{1}{2}(I + N) & \text{vector} \\ W = \frac{1}{2}(I - N) & \text{bivector} \end{cases} \tag{5.16}$$

Noting that $\sigma^2 = 1$, $\sigma e_0 = -e_0\sigma$ and $\sigma e_p = e_p\sigma$ for $p \neq 0$, and then applying the projection operators in either order to a Clifford number $u = \mathcal{C} = (a + A\sigma) + (b + B\sigma)e_0 - i(c + C\sigma)\sigma - i(d + D\sigma)e_0\sigma$, gives:

$$\begin{cases} TW\mathcal{C} = WT\mathcal{C} = & (b + B\sigma)e_0 \\ SW\mathcal{C} = WS\mathcal{C} = & a + A\sigma \\ TV\mathcal{C} = VT\mathcal{C} = & -i(d + D\sigma)e_0\sigma \\ SV\mathcal{C} = VS\mathcal{C} = & -i(c + C\sigma)\sigma \end{cases} \tag{5.17}$$

The S and T operators split into space-like and time-like components (i.e. components without and with unit e_0 respectively). The V and W operators split into components having spatial components which are from odd or even grades. For V that is vectors and trivectors; for W scalars and bivectors.

The S, T splitting operators can also be combined usefully with the boundary data projection (splitting) operators $Q^{\pm}$ as applied to the bivector electromagnetic field $u = \mathcal{F} = \sqrt{\mu}H\sigma - i\sqrt{\epsilon}Ee_0$ in Sect. 3.2.3. The various combinations provide for the splitting of the normal and tangential components of the field into their four constituent parts:

$$\begin{cases} TQ^+\mathcal{F} = Q^+T\mathcal{F} = -i\sqrt{\epsilon}\,(n \cdot E)\,ne_0 \\ SQ^+\mathcal{F} = Q^+S\mathcal{F} = \sqrt{\mu}\,(n \times H) \times n\sigma \\ TQ^-\mathcal{F} = Q^-T\mathcal{F} = -i\sqrt{\epsilon}\,(n \times E) \times ne_0 \\ SQ^-\mathcal{F} = Q^-S\mathcal{F} = \sqrt{\mu}\,(n \cdot H)\,n\sigma \end{cases} \tag{5.18}$$

These relationships come into play in Sect. 6.4.3.

5.3 Frequency Domain

In cases for which an electromagnetic field can be treated a sum of monochromatic components (sinusoidal functions of time), each component can be treated separately as a *phasor* and written:

$$\mathcal{F}(\mathbf{r}, t) = \mathcal{F}_k(\mathbf{r})e^{i\omega t} \tag{5.19}$$

as the product of a Clifford-valued magnitude $\mathcal{F}_k(\mathbf{r})$ and a complex signal of unit amplitude $e^{i\omega t}$ which rotates in a positive angular direction in the complex plane with angular frequency $\omega = 2\pi f > 0$ and wavenumber $k = \omega/c$. The magnitude of the phasor $\mathcal{F}_k(\mathbf{r})$ remains constant in time t but varies with position $\mathbf{r}$, whereas the complex signal does the opposite.

For monochromatic fields, the complex signal can be eliminated and Maxwell's equations can be written in terms the magnitude of phasors alone. This is achieved by expanding the product:

$$\begin{aligned} \mathcal{D}\mathcal{F}_k(\mathbf{r})e^{iwt} &= (\nabla - \tfrac{i}{c}e_0\tfrac{\partial}{\partial t})\mathcal{F}_k(\mathbf{r})e^{iwt} &&= \nabla\mathcal{F}_k(\mathbf{r})e^{iwt} - \tfrac{i}{c}e_0\mathcal{F}_k(\mathbf{r})\tfrac{\partial}{\partial t}e^{iwt} \\ &= \nabla\mathcal{F}_k(\mathbf{r})e^{iwt} - \tfrac{i}{c}e_0\mathcal{F}_k(\mathbf{r})(i\omega)e^{iwt} &&= (\nabla + ke_0)\mathcal{F}_k(\mathbf{r})e^{iwt} \end{aligned} \tag{5.20}$$

The number of variables is effectively reduced from three space and one time, to three space plus a fixed parameter, the wavenumber k. The complex signal can be eliminated from both sides.

Table 5.3 reproduces the electromagnetic quantities given in Table 5.2, converted from temporal to phasor form in the same way as for the field $\mathcal{F}$ in Eq. 5.19. All quantities with a subscript k are the magnitudes of phasors, with the exception of the differential operator $\mathcal{D}_k$.

In the monochromatic frequency domain Maxwell's equations are written in phasor form as:

$$\mathcal{D}_k\mathcal{F}_k = \mathcal{S}_k \tag{5.21}$$

Table 5.3 Electromagnetic phasors in the form of Clifford variables, and quaternions ($\mathfrak{A}$, $\mathfrak{B}$, $\mathfrak{C}$, $\mathfrak{D}$) split into their scalar (a, b, c, d) and vector (A, B, C, D) components

	Clifford	Quaternion							
		$\mathfrak{A}$		$\mathfrak{B}$		$\mathfrak{C}$		$\mathfrak{D}$	
		a	A	b	B	c	C	d	D
Operator	$\mathcal{D}_k = \nabla + ke_0$	0	0	k	0	0	$i\nabla$	0	0
Field	$\mathcal{F}_k = \sqrt{\mu}H_k\sigma - i\sqrt{\epsilon}E_ke_0$	0	$\sqrt{\mu}H_k$	0	0	0	0	0	$-\sqrt{\epsilon}E_k$
Source	$\mathcal{S}_k = \sqrt{\mu}J_k + \frac{i}{\sqrt{\epsilon}}\rho_k e_0$	0	0	$\frac{i}{\sqrt{\epsilon}}\rho_k$	0	0	$i\sqrt{\mu}J_k$	0	0
Potential	$\mathcal{P}_k = \frac{1}{\sqrt{\mu}}A_k + i\sqrt{\epsilon}\phi_k e_0$	0	0	$i\sqrt{\epsilon}\phi_k$	0	0	$\frac{i}{\sqrt{\mu}}A_k$	0	0

The method for the solution of Maxwell's equations described herein is set in the monochromatic frequency domain of Eq. 5.21 rather than the time domain of Eq. 5.5. The method relies on the solution of the homogeneous form of Maxwell's equations:

$$\mathcal{D}_k \mathcal{F}_k(\mathbf{r}) = 0 \tag{5.22}$$

for regions without sources, along with the solution of the inhomogeneous form:

$$\mathcal{D}_k \mathcal{F}_k(\mathbf{r}) = \delta(r - 0) \tag{5.23}$$

for a point source $\mathcal{S}_k(\mathbf{r}) = \delta(r - 0)$ at the origin $r = |\mathbf{r}| = 0$. The two solutions are incorporated into the boundary theorem [30] to produce the electromagnetic (monogenic) Cauchy integral in three dimensions. Details are given in Chap. 7.

Differentiating 5.21 a second time gives:

$$\begin{aligned} \mathcal{D}_k^2 \mathcal{F}_k &= (\nabla + k e_0)(\nabla + k e_0)\mathcal{F}_k \qquad\qquad = (\nabla^2 - k^2)\mathcal{F}_k \\ &= \left(\left(\tfrac{\partial}{\partial x} e_1 + \tfrac{\partial}{\partial y} e_2 + \tfrac{\partial}{\partial z} e_3\right)^2 - k^2\right)\mathcal{F}_k = \left(-\left(\tfrac{\partial^2}{\partial x^2} + \tfrac{\partial^2}{\partial y^2} + \tfrac{\partial^2}{\partial z^2}\right) - k^2\right)\mathcal{F}_k = -H_k \mathcal{F}_k \end{aligned} \tag{5.24}$$

where:

$$H_k = \nabla^2 + k^2 \tag{5.25}$$

is the Helmholtz operator, and:

$$\nabla^2 = \frac{\partial^2}{\partial x^2} + \frac{\partial^2}{\partial y^2} + \frac{\partial^2}{\partial z^2} \tag{5.26}$$

is the Laplacian operator.

The relationship between the operator $\mathcal{D}_k$ and the Helmholtz operator in Eq. 5.24 is of importance because it plays a role in finding the solution of Eq. 5.23, the field for a point source. For details see Sect. 7.2.

5.4 Differential Operators

In the context of electromagnetism a variety of first and second order ordinary and partial differential operators are employed both in the time and frequency domains, and additionally here in both vector and Clifford formalisms. Table 5.4 lists the differential operators as applied in three-dimensional space with or without an additional dimension of time or frequency.The names and symbols adopted for the different operators varies according to source. The Clifford vector differential operator symbolised here as $\mathcal{D}_0$ is referred to most widely as the Dirac operator [31] following its introduction in a quantum mechanical setting [32, 33], although [30] it was introduced by Hamilton almost one century earlier. The operator symbolised as $\mathcal{D}_k$ and

Table 5.4 Differential operators in three or more dimensions

Symbol D	Definition Df	Operand f	Name
∇	$\nabla f = \frac{\partial f}{\partial x}\hat{\mathbf{x}} + \frac{\partial f}{\partial y}\hat{\mathbf{y}} + \frac{\partial f}{\partial z}\hat{\mathbf{z}}$	Scalar	Gradient
$\nabla\cdot$	$\nabla \cdot f = \frac{\partial}{\partial x}(\hat{\mathbf{x}} \cdot f) + \frac{\partial}{\partial y}(\hat{\mathbf{y}} \cdot f) + \frac{\partial}{\partial z}(\hat{\mathbf{z}} \cdot f)$	Vector	Divergence
$\nabla\times$	$\nabla \times f =$ $\frac{\partial}{\partial x}(\hat{\mathbf{x}} \times f) + \frac{\partial}{\partial y}(\hat{\mathbf{y}} \times f) + \frac{\partial}{\partial z}(\hat{\mathbf{z}} \times f)$	Vector	Curl
$\boldsymbol{\nabla}$	$\boldsymbol{\nabla} f = \frac{\partial}{\partial x}(e_1 f) + \frac{\partial}{\partial y}(e_2 f) + \frac{\partial}{\partial z}(e_3 f)$	Clifford	Gradient[a]
$\boldsymbol{\nabla}\cdot$	$\boldsymbol{\nabla} \cdot f =$ $\frac{\partial}{\partial x}(e_1 \cdot f) + \frac{\partial}{\partial y}(e_2 \cdot f) + \frac{\partial}{\partial z}(e_3 \cdot f)$	Clifford	Divergence[b]
$\boldsymbol{\nabla}\times$	$\boldsymbol{\nabla} \times f =$ $\frac{\partial}{\partial x}(e_1 \times f) + \frac{\partial}{\partial y}(e_2 \times f) + \frac{\partial}{\partial z}(e_3 \times f)$	Clifford	Curl[b]
$\mathcal{D}_0$	$\mathcal{D}_0 f = \sum_j \frac{\partial}{\partial x_j}(e_j f)$	Clifford	Dirac
$\mathcal{D}_k$	$\mathcal{D}_k f = \sum_j \frac{\partial}{\partial x_j}(e_j f) + k(e_0 f)$	Clifford	k-Dirac
$\mathcal{D}$	$\mathcal{D} f = \sum_j \frac{\partial}{\partial x_j}(e_j f) - \frac{i}{c}\frac{\partial}{\partial t}(e_0 f)$	Clifford	Hyperbolic Dirac
$\nabla^2, -\boldsymbol{\nabla}^2, -\mathcal{D}_0^2$	$\nabla^2 f = \frac{\partial^2 f}{\partial x^2} + \frac{\partial^2 f}{\partial y^2} + \frac{\partial^2 f}{\partial z^2}$	Scalar	Laplacian
$H_k, -\mathcal{D}_k^2$	$H_k f = \frac{\partial^2 f}{\partial x^2} + \frac{\partial^2 f}{\partial y^2} + \frac{\partial^2 f}{\partial z^2} + k^2 f$	Scalar	Helmholtz
$\Box^2, -\mathcal{D}^2$	$\Box^2 f = \frac{\partial^2 f}{\partial x^2} + \frac{\partial^2 f}{\partial y^2} + \frac{\partial^2 f}{\partial z^2} - \frac{1}{c^2}\frac{\partial^2 f}{\partial t^2}$	Scalar	d'Alembertian

[a] Only if the operand is a scalar
[b] Only if the operand is a vector

variously referred to as the perturbed Dirac operator [34] or the k-Dirac operator [2], reduces to the Dirac operator when $k = 0$.

The k-Dirac operator produces the derivative required when Maxwell's equations are expressed in terms of a monochromatic field in the frequency domain. In the time domain the corresponding operator $\mathcal{D}$ is referred to as the hyperbolic Dirac operator [2].

A special symbol $\boldsymbol{\nabla}$ is adopted in three dimensions only, to distinguish the vector differential operator:

$$\nabla = \frac{\partial}{\partial x}\hat{\mathbf{x}} + \frac{\partial}{\partial y}\hat{\mathbf{y}} + \frac{\partial}{\partial z}\hat{\mathbf{z}} \tag{5.27}$$

in the context of Gibbs's vector calculus [13] from the Cliffordian vector differential operator:

$$\boldsymbol{\nabla} = \frac{\partial}{\partial x}e_1 + \frac{\partial}{\partial y}e_2 + \frac{\partial}{\partial z}e_3 \tag{5.28}$$

in the context of Clifford algebra. The distinction is important because the range of multiplication operations which is supported by each is different.

The vector differential operator ∇ supports operations on scalars by the scalar multiplication of individual components, and operations on vectors by dot and cross multiplications to produce the divergence (a scalar representing the average radial derivative) and the curl (a vector representing the average tangential derivative). The Cliffordian differential operator ∇ supports exactly the same operations. In addition, the Cliffordian differential operator supports Clifford multiplication and all of the variants described in Sect. 2.3.3. For a vector $f = \mathbf{b}$ the Clifford product $\nabla\mathbf{b}$ produces both the curl and the divergence in the form:

$$\nabla\mathbf{b} = (\nabla \times \mathbf{b})\sigma - \nabla \cdot \mathbf{b} \tag{5.29}$$

as in Eq. 1.16 of Sect. 1.5.

The vector differential operators ∇ and ∇ are taken as applying only in three dimensions. This is necessary because as defined the vector cross product applies strictly to three dimensions alone. In three dimensions the Dirac operator $\mathcal{D}_0$ and the Cliffordian vector differential operator ∇ appear identical. However, the Dirac operator does not support the vector cross product, and that allows it to operate properly in all dimensions without special cases.

5.4.1 Leibniz Product Rule

The Clifford algebra derivative of a product of a pair of functions can be cast in a coordinate-free form in terms of the areal derivative defined by Pompeiu [35, 36]. For two functions g and f at a point $\mathbf{r}$ in a region Ω enclosed by a boundary Σ the derivative is given by [37]:

$$D(g|f)(\mathbf{r}) = \lim_{\Omega\downarrow\mathbf{r}} \frac{\int_\Sigma g(\mathbf{s})n(\mathbf{s})f(\mathbf{s})\,d\sigma(\mathbf{s})}{\int_\Omega 1\,d\mathbf{s}} \tag{5.30}$$

where $n(\mathbf{s})$ is the outward normal to the boundary of the region, $\mathbf{s}$ is a point on the boundary, $d\sigma(\mathbf{s})$ is the elemental measure of boundary, and $d\mathbf{s}$ is the elemental measure of the interior. The limit indicates that the region shrinks onto the point $\mathbf{r}$.

The coordinate-free derivative of a single function f can be calculated by setting the function $g = 1$:

$$D(1|f)(\mathbf{r}) = \lim_{\Omega\downarrow\mathbf{r}} \frac{\int_\Sigma n(\mathbf{s})f(\mathbf{s})\,d\sigma(\mathbf{s})}{\int_\Omega 1\,d\mathbf{s}} \tag{5.31}$$

When any particular coordinate system is chosen the Eqs. 5.30 and 5.31 for the coordinate-free derivative can be cast in terms of the basis vectors of the system. For example, consider the case in two dimensions for Cartesian coordinates with primal basis vectors e_1, e_2 and a circular region of radius Δr having normal $n(\mathbf{s}) = (\cos\theta)e_1 + (\sin\theta)e_2$ as shown in Fig. 5.1, and assume that the function $f(\mathbf{r}) = f(x, y) = f_x e_1 + f_y e_2$ is represented by a Taylor series:

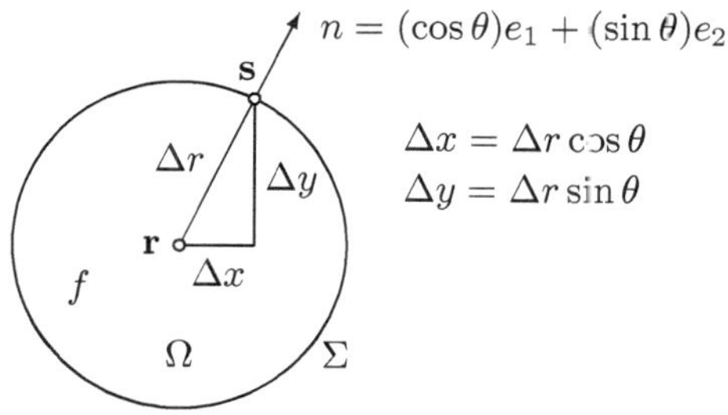

Fig. 5.1 Circular region Ω with boundary Σ containing Clifford-valued function f

$$\begin{aligned} f(\mathbf{s}) =& f(x+\Delta x, y+\Delta y) = (f_x + \Delta x \tfrac{\partial f_x}{\partial x} + \Delta y \tfrac{\partial f_x}{\partial y})e_1 \\ &+ (f_y + \Delta x \tfrac{\partial f_y}{\partial x} + \Delta y \tfrac{\partial f_y}{\partial y})e_2 + \mathcal{O}(\Delta r^2) \end{aligned} \tag{5.32}$$

If the second and higher order derivatives are bounded and if the circular region is small enough, then the first few terms of the Taylor series are often a good approximation to the value of the function. This is normally the case if the function f is well behaved (i.e. does not contain a singularity).

Retaining the constant and linear terms only, the product $n(\mathbf{s})f(\mathbf{s})$ gives:

$$\begin{aligned} nf &= [(\cos\theta)e_1 + (\sin\theta)e_2]\left[\left(f_x + \Delta x \tfrac{\partial f_x}{\partial x} + \Delta y \tfrac{\partial f_x}{\partial y}\right)e_1 + \left(f_y + \Delta x \tfrac{\partial f_y}{\partial x} + \Delta y \tfrac{\partial f_y}{\partial y}\right)e_2\right] \\ &= \left[(\cos\theta)\left(f_y + \Delta x \tfrac{\partial f_y}{\partial x} + \Delta y \tfrac{\partial f_y}{\partial y}\right) - (\sin\theta)\left(f_x + \Delta x \tfrac{\partial f_x}{\partial x} + \Delta y \tfrac{\partial f_x}{\partial y}\right)\right]e_1e_2 \\ &\quad - \left[(\cos\theta)\left(f_x + \Delta x \tfrac{\partial f_x}{\partial x} + \Delta y \tfrac{\partial f_x}{\partial y}\right) + (\sin\theta)\left(f_y + \Delta x \tfrac{\partial f_y}{\partial x} + \Delta y \tfrac{\partial f_y}{\partial y}\right)\right] \end{aligned} \tag{5.33}$$

So that:

$$\begin{aligned} D(1|f)(\mathbf{r}) &= \lim_{\Delta r\to 0} \frac{\int_\Sigma nf\, d\sigma(\mathbf{s})}{\int_\Omega 1\, d\mathbf{s}} = \lim_{\Delta r\to 0} \frac{\int_{\theta=0}^{2\pi} nf\, \Delta r\, d\theta}{\pi(\Delta r)^2} \\ &= \lim_{\Delta r\to 0} \frac{1}{\pi\Delta r}\int_{\theta=0}^{2\pi} \Delta r\left[(\cos^2\theta)\tfrac{\partial f_y}{\partial x} - (\sin^2\theta)\tfrac{\partial f_x}{\partial y}\right]e_1e_2 \\ &\quad -\Delta r\left[(\cos^2\theta)\tfrac{\partial f_x}{\partial x} + (\sin^2\theta)\tfrac{\partial f_y}{\partial y}\right]d\theta \end{aligned} \tag{5.34}$$

where other terms containing $\cos\theta$, $\sin\theta$ and $\sin 2\theta$ have been omitted from Eq. 5.34 because their integrals over the range from $\theta = 0$ to 2π are known in advance to give a result of zero. Evaluating the remaining integrals gives:

$$\begin{aligned} D(1|f)(\mathbf{r}) &= \tfrac{1}{\pi}\left\{\left[\pi\tfrac{\partial f_y}{\partial x} - \pi\tfrac{\partial f_x}{\partial y}\right]e_1e_2 - \left[\pi\tfrac{\partial f_x}{\partial x} + \pi\tfrac{\partial f_y}{\partial y}\right]\right\} = \left[\tfrac{\partial f_y}{\partial x} - \tfrac{\partial f_x}{\partial y}\right]e_1e_2 - \left[\tfrac{\partial f_x}{\partial x} + \tfrac{\partial f_y}{\partial y}\right] \\ &= \left(\tfrac{\partial}{\partial x}e_1 + \tfrac{\partial}{\partial y}e_2\right)(f_xe_1 + f_ye_2) \qquad\qquad = \left(\tfrac{\partial}{\partial x}e_1 + \tfrac{\partial}{\partial y}e_2\right)f \end{aligned} \tag{5.35}$$

It is recognised from Eq. 5.35 that the derivative $D(1|f)$ is the same as the Clifford product $\mathcal{D}_0 f$, where $\mathcal{D}_0$ is the Dirac operator. By similar reasoning, if the function g is also well approximated by a Taylor series then $D(g|1)$ is the same as the Clifford product $g\mathcal{D}_0$, and furthermore:

$$D(g|f) = D(g|1)f + gD(1|f) \tag{5.36}$$

Equation 5.36 serves as the Leibniz rule for the Clifford differentiation of a product of two Clifford-valued functions. Written in terms of the Dirac operator, the product rule takes the form:

$$D(g|f) = \mathcal{D}_0(gf) = \sum_p \frac{\partial}{\partial x_p}(ge_p f) = (g\mathcal{D}_0)f + g(\mathcal{D}_0 f) \tag{5.37}$$

The Clifford product $\mathcal{D}_0 f$ represents differentiation of the function f by applying the differential operator $\mathcal{D}_0$ from the left hand side. The Clifford product $g\mathcal{D}_0$ represents differentiation of the function g by applying the differential operator $\mathcal{D}_0$ from the right hand side.

The key point to note here is that the primal unit e_p which is part of the Dirac operator $\mathcal{D}_0$ plays the role of the normal vector in Eq. 5.30 and sits *between* the two functions g and f. For example, in three dimensions:

$$\begin{aligned}
\mathcal{D}_0(gf) &= \sum_{p=1}^{3} \frac{\partial}{\partial x_p}(ge_p f) \\
&= \frac{\partial}{\partial x_1}(ge_1 f) + \frac{\partial}{\partial x_2}(ge_2 f) + \frac{\partial}{\partial x_3}(ge_3 f) \\
&= \left(\frac{\partial g}{\partial x_1}e_1 f + ge_1\frac{\partial f}{\partial x_1}\right) + \left(\frac{\partial g}{\partial x_2}e_2 f + ge_2\frac{\partial f}{\partial x_2}\right) + \left(\frac{\partial g}{\partial x_3}e_3 f + ge_3\frac{\partial f}{\partial x_3}\right) \\
&= \left(\frac{\partial g}{\partial x_1}e_1 + \frac{\partial g}{\partial x_2}e_2 + \frac{\partial g}{\partial x_3}e_3\right) f + g\left(e_1\frac{\partial f}{\partial x_1} + e_2\frac{\partial f}{\partial x_2} + e_3\frac{\partial f}{\partial x_3}\right) \\
&= \left(\reflectbox{$\frac{\partial g}{\partial x_1}$}e_1 + \reflectbox{$\frac{\partial g}{\partial x_2}$}e_2 + \reflectbox{$\frac{\partial g}{\partial x_3}$}e_3\right) f + g\left(e_1\frac{\partial f}{\partial x_1} + e_2\frac{\partial f}{\partial x_2} + e_3\frac{\partial f}{\partial x_3}\right) \\
&= \left[\reflectbox{g}\left(\reflectbox{$\frac{\partial}{\partial x_1}$}e_1 + \reflectbox{$\frac{\partial}{\partial x_2}$}e_2 + \reflectbox{$\frac{\partial}{\partial x_3}$}e_3\right)\right] f + g\left[\left(e_1\frac{\partial}{\partial x_1} + e_2\frac{\partial}{\partial x_2} + e_3\frac{\partial}{\partial x_3}\right) f\right] \\
&= \reflectbox{g}{}_0\reflectbox{$\mathcal{D}$} f + g\mathcal{D}_0 f \\
&= (g\mathcal{D}_0)f + g(\mathcal{D}_0 f)
\end{aligned} \tag{5.38}$$

The reversed symbols $\reflectbox{$\partial$}$ and ${}_0\reflectbox{$\mathcal{D}$}$ are used temporarily to emphasise that those operators are applied from the right hand side onto operands further towards the left. Normally the same operation is indicated by parentheses without reversing the symbols, as in the last line of Eq. 5.38.

5.5 Exercises

Q1. Show that $-\nabla^2$ with the minus sign is the same as the Laplacian $+\nabla^2$ with the plus sign.

Q2. Show that $\mathcal{D}^2\mathcal{P}$ separates according to the independence of quaternion units into wave equations for both the scalar electric and vector magnetic potentials.

Q3. Show that $\mathcal{D}^2\mathcal{F}$ separates according to the independence of quaternion units into wave equations for both the electric and magnetic fields.

Q4. Verify that $\sigma^2 = +1$.

Q5. Verify that $\sigma e_0 = -e_0\sigma$.

Q6. Verify that $\sigma e_p = e_p\sigma$ for $p > 0$.

Q7. Verify that M and N in Eqs. 5.13 and 5.14 are reflection operators.

Q8. Show that the reflection operators M and N are commutative.

Q9. Show that the projection operators S and T are each commutative with the operators V and W.

Q10. Show that the projection operators S and T are each commutative with the boundary data projection operators Q^+ and Q^-.

Q11. Verify that applying projection operators S, T, V and W to Eq. 5.6 as in Eq. 5.17 produces Eq. 5.7.

Q12. Verify that applying projection operators S, T, V and W to Eq. 5.9 as in Eq. 5.17 produces Eq. 5.10.

Q13. Verify that applying projection operators S, T, Q^+ and Q^- to the electromagnetic field $\mathcal{F} = \sqrt{\mu}H\sigma - i\sqrt{\epsilon}Ee_0$ produces Eq. 5.18.

Q14. Determine the relationship between the energy in an electromagnetic wave and the scalar product of the electromagnetic field $(\mathcal{F}, \mathcal{F}^*)$, where $\mathcal{F}^*$ is the complex conjugate of $\mathcal{F}$.

Q15. Show that any second order terms retained in the Taylor series 5.32 for use in the product nf of Eq. 5.33 have negligible effect as the region Ω shrinks onto the point $\mathbf{r}$.

Q16. Starting with Eq. 5.30, show in two dimensions using Cartesian coordinates that the areal derivative $D(g|1)$ is the same as the Clifford expression $g\mathcal{D}_2$.

Q17. Starting with the areal derivative in Eq. 5.30, using two Taylor series and Cartesian coordinates in two dimensions, verify the product rule of Eq. 5.36.

Part II
Formulation

Chapter 6
Scattering

Figure 6.1 shows electromagnetic radiation from a distant source ★ which encounters an object Ω^+ having values of electric permittivity ϵ and magnetic permeability μ different from the surrounding space, Ω^-. The radiation is in part excluded and redirected away from the object and is in part admitted and borne into the object.

The wave $f^\star$ arriving from the source is called the *incident* wave. The excluded component of the wave f^- is called the *scattered* or *reflected* wave, and the admitted component f^+ is called the *transmitted* wave. The term *reflected* is used more often in cases where the overall shape of the scattered wave strongly resembles that of the incident wave.

The scattering of electromagnetic radiation from various objects occurs frequently in electrical engineering. In some cases, such as radar, the object may be an aeroplane or a ship and exhibits the properties of a metallic conductor. In other cases, such as optics or opto-electronics, the object may be a lens or a waveguide, and has the properties of a lossless dielectric. In applications where it is intended to hide a conducting object from detection by radar, a lossy dielectric may be used as a surface layer to absorb the radiation admitted and to reduce the amount of, or eliminate, the radiation scattered.

In most of these applications the different objects involved have material properties μ, ϵ which take constant and uniform values within well defined regions of space, and which change abruptly across interfaces formed where the boundaries of two such regions meet. Within each and every of all such regions Maxwell's equations apply in the homogeneous form ($\mathcal{D}\mathcal{F} = 0$ in the time domain or $\mathcal{D}_k\mathcal{F}_k = 0$ in the frequency domain) because the problem is posed in such a way that the sources (although not their effects) are effectively excluded.

At interfaces between regions, Maxwell's equations are replaced by simpler equations. These relate to one another the normal and tangential fields immediately on either side of the interface, in terms of the values of the material properties on the two sides.

A. Seagar, *Application of Geometric Algebra to Electromagnetic Scattering*,
DOI 10.1007/978-981-10-0089-8_6

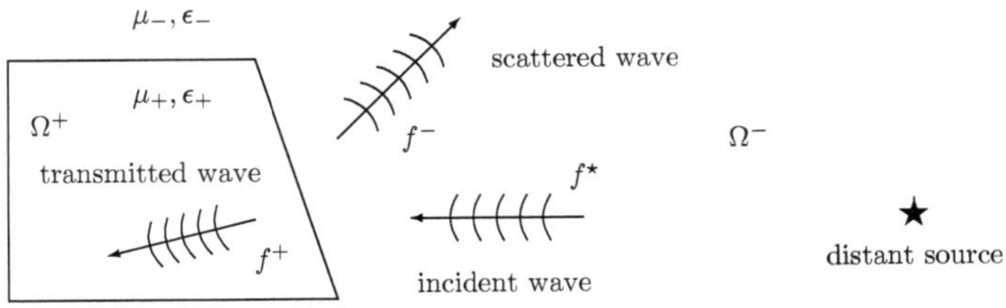

Fig. 6.1 Radiation from distant source ★ partially borne into and scattered away from object Ω^+

In order to design systems which interact with electromagnetic fields in whatever ways are deemed useful involves determining the scattered and admitted fields f^- and f^+, from knowledge of the incident field $f^\star$, the material properties of all regions, and a geometric specification of their boundaries.

6.1 Examples

The simplest practical example of scattering in electrical engineering is that for a transmission line where the characteristic impedance changes abruptly from one value Z_- to another Z_+ as shown in Fig. 6.2a. The amplitude of the reflected and transmitted waves is governed by the reflection coefficient Γ and the transmission coefficient τ:

$$\begin{cases} \Gamma = \dfrac{Z_+ - Z_-}{Z_+ + Z_-} \\ \tau = \dfrac{2Z_+}{Z_+ + Z_-} \end{cases} \tag{6.1}$$

For unguided waves propagating in three dimensions the corresponding case involves a linearly polarised plane wave scattering from an infinite planar surface perpendicular to the direction of motion, as shown in Fig. 6.2b. In this case the

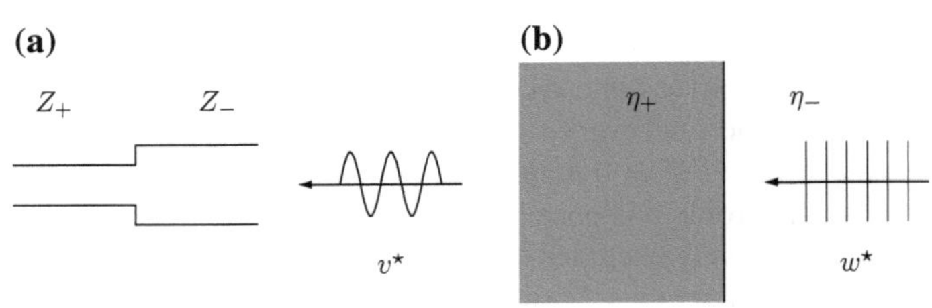

Fig. 6.2 One-dimensional scattering (**a**) of voltage $v^\star$ from discontinuity in characteristic impedance Z on transmission line, and (**b**) of plane electromagnetic wave $w^\star$ from discontinuity in intrinsic impedance η at planar surface

reflection and transmission coefficients are governed by the material properties on each side of the interface at the surface, expressed in the form of intrinsic impedances $\eta_\pm = \sqrt{\mu_\pm/\varepsilon_\pm}$:

$$\begin{cases} \Gamma = \dfrac{\eta_+ - \eta_-}{\eta_+ + \eta_-} \\ \tau = \dfrac{2\eta_+}{\eta_+ + \eta_-} \end{cases} \tag{6.2}$$

Two special cases are of interest. Firstly is the case when $\eta_+/\eta_- = 1$. There is no reflection ($\Gamma = 0$) and complete transmission ($\tau = 1$). Secondly is the case when $\eta_+/\eta_- = 0$. Then there is complete reflection ($\Gamma = -1$) and no transmission ($\tau = 0$). The negative sign indicates that the reflected wave is inverted.

Note in the definition of the intrinsic impedance that $\varepsilon_\pm = \epsilon_\pm - i\sigma_\pm/\omega$ is the effective permittivity for a sinusoidal field of angular frequency ω. This effective permittivity contains an imaginary component which is related to the electrical conductivity $\sigma_\pm$. When the electrical conductivity is low then $\sigma_\pm \ll \omega\epsilon_\pm$ and displacement current flow dominates Ohmic current flow. The conductive component can be safely ignored and the effective permittivity $\varepsilon_\pm \approx \epsilon_\pm$ is no different from that for a lossless dielectric. However, for a good conductor $\sigma_\pm \gg \omega\epsilon_\pm$ and Ohmic current flow dominates. In that case the conductive component cannot be ignored and the effective permittivity $\varepsilon_\pm \approx -i\sigma_\pm/\omega$ must be used instead of $\epsilon_\pm$.

6.2 Single Regions

In two dimensions Cauchy showed [8][1] that the solution for a complex-valued function $f(z)$ satisfying the homogeneous equation $df/dz = 0$ within a closed boundary, is given directly and explicitly in terms of the Cauchy integral of the trace (value) of the function on the boundary. The method was later generalised in multiple dimensions by McIntosh [10] for Clifford-valued functions $f(\mathbf{r})$ satisfying the homogeneous equation $\mathcal{D}_k f = 0$ within Lipschitz domains. This is exactly the case for the monochromatic electromagnetic field $f = \mathcal{F}_k$ in a region without sources.

In two dimensions the function $f(z)$ is described as an *analytic* function of a complex variable z. In the multi-dimensional case the function $f(\mathbf{r})$ is described either as a *monogenic* or as a *k-monogenic* function of a Clifford vector variable $\mathbf{r}$. The term *monogenic* without the k is used in the case of static fields, i.e. when $k = 0$.

The electromagnetic field in a region without sources behaves like a k-monogenic function of a Clifford variable, and is determined directly and explicitly from the multi-dimensional Cauchy integral of the trace of the field on the boundary of the region.

Details of the particular form of the Cauchy integral required in three dimensions is given in Sect. 7.3.2.

[1] Equation 6.

6.3 Multiple Regions

In the case where there are multiple regions with material properties taking different values, solution of the scattering problem involves satisfying simultaneously two conditions:

- Maxwell's homogeneous equations within each and every region, and
- the boundary conditions which relate to one another the fields on adjacent sides of all interfaces.

The first condition is met by the imposition of Maxwell's homogeneous equations in the form of two projection operators called *Hardy* projections. The particular Hardy projections required have been derived from the multi-dimensional Cauchy integral by Axelsson [2] for the specific purpose of relating a joint function at the interface to the fields immediately on either side. The Cauchy integral is described in Chap. 7, and the Hardy projections are described in Chap. 8.

The second condition is met by imposing equations derived from the boundary conditions, cast in the form of two boundary data projection operators. Each of the two projection operators relates one half of the unknown field on the boundary to one half of the incident field on the boundary. These operators are described in Sect. 6.4.

6.3.1 Extreme Cases

Of all possible values for material properties, two cases stand out as at the opposite extremes of a continuum. Both of these cases are of practical interest, and they are both easier to solve than any other.

The first case is that where all radiation is scattered. The electrical conductivity inside the object is effectively infinite. This extreme value of material property dominates the solution, and the values of the electric permittivity and magnetic permeability within the object play no role. The nett (total) electric and magnetic fields on either side of the interface at the object's surface (i.e. incident plus scattered outside, and transmitted inside) are wholly discontinuous, with those on the inside being identically zero. Only two fields are involved, the incident field and the scattered field.

The context in which the first case if of interest is for metallic objects, most of which have a value of electric conductivity high enough to be taken as infinite. Applications are in the detection of distant objects by radar, and in the design of antennas and systems of antennas.

The second case is that where the electrical conductivity inside the object is at the other extreme, zero, and the electric permittivity and magnetic permeability govern the solution. Their particular values inside are taken as equal to those outside. The object is perfectly transparent to electromagnetic radiation, although it may be possible to detect it using other means. The incident wave is neither scattered nor absorbed in any way. The transmitted wave can in no way be distinguished from what would be measured in the region occupied by the object should the object be removed. The nett electric and magnetic fields on either side of the interface at the object's surface are wholly continuous, those immediately inside being identical to those immediately outside. Again only two fields are involved, this time the incident field and the transmitted field.

The application in which the second case is of interest is for the extension into the far field of a radiation pattern measured (perhaps only partially) near an antenna. In this case the source is inside the object, and the object itself is defined by the surface over which the field is measured. The values of electric permittivity and magnetic permeability are everywhere the same.

In both of these cases each of the two conditions which govern the solution at the surface of the object are represented by projection operators, and have a geometrical interpretation as such. The solution itself has a geometrical interpretation as the intersection of two straight lines, one for each condition, and can be found by algebraically implementing geometrical operations as in Sect. 3.5. The full description of these cases and their solutions are given in Sects. 9.1 and 9.2.

Other cases, such as for lossless and lossy dielectrics, fall somewhere on the continuum in between the two extremes. These cases are more complicated since they require three fields to be matched on the interface at the object's surface rather than just two. The full description of these more general cases and their solution is given in Sect. 9.3. The geometrical interpretation of the solution is that of a vector linking one line to another.

6.3.2 Components of Solutions

The components required in the construction of the solutions are listed in Table 6.1. The Hardy projection operators are more complicated than the boundary data projection operators because the Hardy projections require operations globally over entire surfaces whereas the boundary data projections require operations only locally at individual points.

Table 6.1 Components of solution

Component		Perfect reflection	Perfect transmission	Partial transmission and reflection
1a	Boundary conditions	Sect. 6.4	Sect. 6.4	Sect. 6.4
1b	Boundary data projection operators	Sect. 6.4.1	Sect. 6.4.2	Sect. 6.4.3
2a	Maxwell's equations	Sect. 5.3	Sect. 5.3	Sect. 5.3
2b	Hardy projection operators	Sect. 8.1	Sect. 8.2	Sect. 8.3
3a	Formulation	Sect. 9.1	Sect. 9.2	Sects. 9.3, 9.3.1
3b	Solution	Sect. 9.1	Sect. 9.2	Sect. 9.3.2

Table 6.2 Stages in construction of the Cauchy integral

	Stage	Context	Section
1	Boundary theorem	Fundamental theorem of calculus	Sect. 7.1
2	Fundamental solution	Point source	Sect. 7.2
3	Reproducing formula	In one region	Sect. 7.3.1
4	Cauchy extension	Into two adjacent regions	Sect. 7.3.2
5	Cauchy integral	On interface between regions	Sect. 7.4.1

The Hardy projection operators are constructed from the Cauchy integral. That integral is developed in stages from the boundary theorem as listed in Table 6.2.

6.4 Boundary Conditions

Electromagnetic boundary field problems fall naturally into three different cases, as shown in Table 6.3, according to the behaviour of the field at the interface between regions:

- perfect reflection,
- perfect transmission, or
- partial transmission and reflection.

Table 6.3 Three cases of electromagnetic boundary field problems in the notation of vector calculus

Case	Fields	Boundary conditions
Perfect reflection[a] $\epsilon_+ = \infty,\ \mu_+ = 0$ $\sigma_+ = \infty$	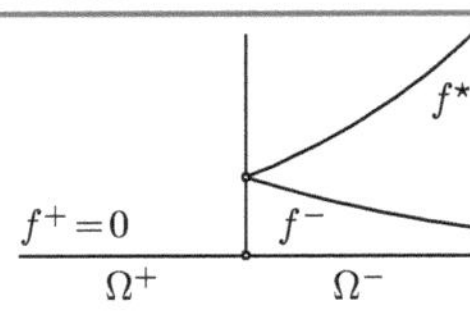	$\epsilon_- \mathbf{n} \cdot [\mathbf{E}_\star + \mathbf{E}_-] = \rho_s$ $\mathbf{n} \times [\mathbf{H}_\star + \mathbf{H}_-] = \mathbf{J}_s$ $\mathbf{n} \times [\mathbf{E}_\star + \mathbf{E}_-] = 0$ $\mu_- \mathbf{n} \cdot [\mathbf{H}_\star + \mathbf{H}_-] = 0$
Perfect transmission[a] $\epsilon_+ = \epsilon_-,\ \mu_+ = \mu_-$ $\sigma_+ = \sigma_-$	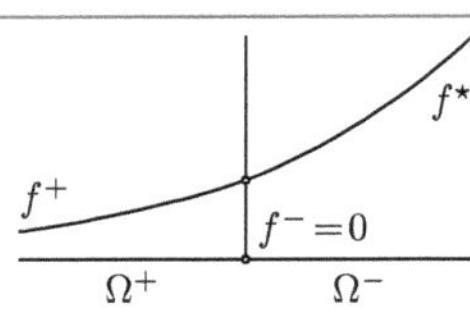	$\mathbf{n} \cdot \mathbf{E}_\star - \mathbf{n} \cdot \mathbf{E}_+ = 0$ $\mathbf{n} \times \mathbf{H}_\star - \mathbf{n} \times \mathbf{H}_+ = 0$ $\mathbf{n} \times \mathbf{E}_\star - \mathbf{n} \times \mathbf{E}_+ = 0$ $\mathbf{n} \cdot \mathbf{H}_\star - \mathbf{n} \cdot \mathbf{H}_+ = 0$
Partial transmission and reflection	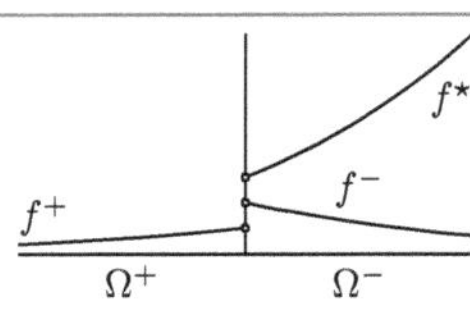	$\epsilon_- \mathbf{n} \cdot [\mathbf{E}_\star + \mathbf{E}_-] - \epsilon_+ \mathbf{n} \cdot \mathbf{E}_+ = 0$ $\mathbf{n} \times [\mathbf{H}_\star + \mathbf{H}_-] - \mathbf{n} \times \mathbf{H}_+ = 0$ $\mathbf{n} \times [\mathbf{E}_\star + \mathbf{E}_-] - \mathbf{n} \times \mathbf{E}_+ = 0$ $\mu_- \mathbf{n} \cdot [\mathbf{H}_\star + \mathbf{H}_-] - \mu_+ \mathbf{n} \cdot \mathbf{H}_+ = 0$

[a]Here $\sigma_\pm$ is the electrical conductivity

At the interface the normal **n** is in the same direction as the outward normal to the boundary of the region Ω^+. In the case of perfect reflection, ρ_s and $\mathbf{J}_s$ are respectively the surface electric charge density and surface electric current density induced on the interface.

The equations for the boundary conditions involving dot products arise from the requirement (as embodied in by Gausses law) that the normal component of the flux density emerging on both sides of the interface is balanced by the surface charge density present on the interface. Similarly, the equations for the boundary conditions involving cross products arise from the requirement that the tangential component of the field circulating around both sides of a finite section of the interface is balanced by the surface current density flowing within the interface in a direction perpendicular to that component of field.

For electric fields the surface electric charge density ρ_s may take non-zero values but the surface magnetic current density is identically zero. Contrarily, for magnetic fields the surface magnetic charge density is identically zero while the surface electric current density $\mathbf{J}_s$ may be otherwise.

In the case of perfect reflection, the values of the surface electric charge and current densities induced on the interface are indeed non-zero. In other cases, there may be induced non-zero values of volumetric charge or current densities, but their volumetric value at the interface is finite. Once integrated over the infinitesimal thickness of the interface their effect as a surface density vanishes.

6.4.1 Perfect Reflection

In the case of perfect reflection, the surface current density $\mathbf{J}_s$ and charge density ρ_s are not known at the outset. Attempting as part of the solution to use the first two equations in the upper section of Table 6.3 introduces two new unknowns which need to be eliminated. That can be achieved by the use of two suitable equations. The simplest approach is to use the very same two equations, by discarding them entirely. Once the solution is found the two equations so discarded can be used to calculate the induced surface current and charge density, if those are indeed required.

Having eliminated the first two equations now leaves only:

$$\mathbf{n} \times [\mathbf{E}_\star + \mathbf{E}_-] = 0 \tag{6.3}$$

$$\mu_- \mathbf{n} \cdot [\mathbf{H}_\star + \mathbf{H}_-] = 0 \tag{6.4}$$

A Clifford version of these equations is constructed by multiplying Eq. 6.4 with $1/\sqrt{\mu_-}$ on the left and with $n\sigma$ on the right, by multiplying Eq. 6.3 with $-i\sqrt{\epsilon_-}$ on the left and by vector cross multiplying with ne_0 on the right, and adding the two results together:

$$[\sqrt{\mu_-}\,(n \cdot H_\star)n\sigma - i\sqrt{\epsilon_-}\,(n \times E_\star) \times ne_0] + [\sqrt{\mu_-}\,(n \cdot H_-)n\sigma - i\sqrt{\epsilon_-}\,(n \times E_-) \times ne_0] = 0 \tag{6.5}$$

All vectors have been replaced by their plain typeface Clifford equivalents, as in Table 5.1.

Reference to Eq. 3.13 shows this Clifford-valued boundary condition can be rewritten more compactly as:

$$Q^- f^\star + Q^- f^- = 0 \tag{6.6}$$

where Q^- is one of the boundary projection operators as in Sects. 3.2.1 and 3.2.3, and where:

$$\begin{cases} f^\star = \sqrt{\mu_-}\, H_\star\sigma - i\sqrt{\epsilon_-}\, E_\star e_0 \\ f^- = \sqrt{\mu_-}\, H_-\sigma - i\sqrt{\epsilon_-}\, E_- e_0 \end{cases} \tag{6.7}$$

are the bivector Clifford forms of the electromagnetic fields, as in Sect. 5.1.2 or 5.3.

Although there is now only one boundary Eq. 6.6 it can be split at any time into the two original components through the use of the space-like S and time-like T projection operators described in Sect. 5.2.

The value of the incident field $f^\star$ is known, so that the equation:

$$g = -Q^- f^\star \tag{6.8}$$

can be calculated directly and taken as data for any particular problem. The boundary condition (6.6) in Clifford form is then simply:

$$Q^- f^- = g \tag{6.9}$$

The equation can be read to state *"certain components of the reflected field on the left match certain components of the incident field on the right"*. The solution based on this boundary condition is described in Sect. 9.1.

6.4.2 Perfect Transmission

In the case of perfect transmission, either the first two or the last two of the equations in the middle section of Table 6.3 can be used in the solution. If the last two are used, the solution proceeds in much the same way as described for the case of perfect reflection (see Sect. 6.4.1) with the introduction of the Q^- boundary projection operator. If the first two equations are used:

$$\mathbf{n} \cdot [\mathbf{E}_\star - \mathbf{E}_+] = 0 \tag{6.10}$$

$$\mathbf{n} \times [\mathbf{H}_\star - \mathbf{H}_+] = 0 \tag{6.11}$$

it is the boundary projection operator Q^+ which must be introduced instead.

A Clifford version of the boundary conditions is constructed by multiplying Eq. 6.10 with $-i\sqrt{\epsilon_\pm}$ on the left and with ne_0 on the right, by multiplying Eq. 6.11

with $\sqrt{\mu_\pm}$ on the left and by vector cross multiplying with $n\sigma$ on the right, and adding the two results together:

$$\begin{aligned}&[\sqrt{\mu_\pm}\,(n \times H_\star) \times n\sigma - i\sqrt{\epsilon_\pm}\,(n \cdot E_\star)ne_0]\\ &\quad - [\sqrt{\mu_\pm}\,(n \times H_+) \times n\sigma - i\sqrt{\epsilon_\pm}\,(n \cdot E_+)ne_0] = 0\end{aligned} \tag{6.12}$$

Here the material properties $\epsilon_\pm = \epsilon_+ = \epsilon_-$ and $\mu_\pm = \mu_+ = \mu_-$ are the same on both sides of the interface.

Reference to Eq. 3.12 shows this Clifford-valued boundary condition can be rewritten more compactly as:

$$Q^+ f^\star - Q^+ f^+ = 0 \tag{6.13}$$

where Q^+ is one of the projection operators as in Sects. 3.2.1 and 3.2.3, and where:

$$\begin{cases} f^\star &= \sqrt{\mu_\pm}\, H_\star\sigma \; - i\sqrt{\epsilon_\pm}\, E_\star e_0 \\ f^+ &= \sqrt{\mu_\pm}\, H_+\sigma - i\sqrt{\epsilon_\pm}\, E_+ e_0 \end{cases} \tag{6.14}$$

are the bivector Clifford forms of the electromagnetic fields, as in Sect. 5.1.2 or 5.3.

Although there is now only one boundary Eq. 6.13 it can be split at any time into the two original components through the use of the space-like S and time-like T projection operators described in Sect. 5.2.

The value of the incident field $f^\star$ is known, so that the equation:

$$h = Q^+ f^\star \tag{6.15}$$

can be calculated directly and taken as data for any particular problem. The boundary condition 6.13 in Clifford form is then simply:

$$Q^+ f^+ = h \tag{6.16}$$

The equation can be read to state *"certain components of the transmitted field on the left match certain components of the incident field on the right"*. The solution based on this boundary condition is described in Sect. 9.2.

6.4.3 Partial Transmission and Reflection

In the case that part of the electromagnetic wave is transmitted and part is reflected, all four equations in the lower section of Table 6.3:

$$\begin{cases} \epsilon_-\mathbf{n} \cdot [\,\mathbf{E}_\star + \mathbf{E}_-\,] \; - \epsilon_+\mathbf{n} \cdot \mathbf{E}_+ = 0 \\ \mathbf{n} \times [\,\mathbf{H}_\star + \mathbf{H}_-\,] \; - \mathbf{n} \times \mathbf{H}_+ = 0 \\ \mathbf{n} \times [\,\mathbf{E}_\star + \mathbf{E}_-\,] \; - \mathbf{n} \times \mathbf{E}_+ = 0 \\ \mu_-\mathbf{n} \cdot [\,\mathbf{H}_\star + \mathbf{H}_-\,] - \mu_+\mathbf{n} \cdot \mathbf{H}_+ = 0 \end{cases} \tag{6.17}$$

are used. As in Sects. 6.4.1 and 6.4.2 the aim is to bind these into a single Clifford-valued equation, without destroying any information, using the projection operators Q^- and Q^+ along with bivector forms of all of the electromagnetic fields from Eqs. 6.7 and 6.14:

$$\begin{cases} f^\star = \sqrt{\mu_-}\,H_\star\sigma - i\sqrt{\epsilon_-}\,E_\star e_0 \\ f^- = \sqrt{\mu_-}\,H_-\sigma - i\sqrt{\epsilon_-}\,E_- e_0 \\ f^+ = \sqrt{\mu_+}\,H_+\sigma - i\sqrt{\epsilon_+}\,E_+ e_0 \end{cases} \tag{6.18}$$

With four vector style boundary conditions rather than two, it is now necessary to use in conjunction with the $Q^\pm$ boundary projection operators two more, in particular the space-like and time-like projection operators S and T described in Sect. 5.2.

The procedure for constructing the Clifford version of the boundary conditions follows the same steps of multiplying Eq. 6.17 by certain values and adding with the aim of combining all separate electric and magnetic field components from Eq. 6.17 into the Clifford forms of Eq. 6.18.

Firstly multiply Eq. 6.17 in the order listed by $1/i\sqrt{\epsilon_-}$, $\sqrt{\mu_+}$, $-i\sqrt{\epsilon_+}$ and $1/\sqrt{\mu_-}$ respectively on the left, and multiply (with cross product $\times$ as indicated) by ne_0, $\times n\sigma$, $\times ne_0$ and $n\sigma$ respectively on the right:

$$\begin{cases} \{-i\sqrt{\epsilon_-}(n\cdot[E_\star + E_-])\,ne_0\} - \sqrt{\epsilon_+/\epsilon_-}\,\{-i\sqrt{\epsilon_+}(n\cdot E_+)\,ne_0\} = 0 \\ \sqrt{\mu_+/\mu_-}\,\{\sqrt{\mu_-}(n\times[H_\star + H_-])\times n\sigma\} - \{\sqrt{\mu_+}(n\times H_+)\times n\sigma\} = 0 \\ \sqrt{\epsilon_+/\epsilon_-}\,\{-i\sqrt{\epsilon_-}(n\times[E_\star + E_-])\times ne_0\} - \{-i\sqrt{\epsilon_+}(n\times E_+)\times ne_0\} = 0 \\ \{\sqrt{\mu_-}(n\cdot[H_\star + H_-])\,n\sigma\} - \sqrt{\mu_+/\mu_-}\,\{\sqrt{\mu_+}(n\cdot H_+)\,n\sigma\} = 0 \end{cases} \tag{6.19}$$

Introducing for notational convenience the contrast factors α, β, which are related to the material properties by:

$$\begin{cases} \alpha = \sqrt{\mu_+/\mu_-} \\ \beta = \sqrt{\epsilon_+/\epsilon_-} \end{cases} \tag{6.20}$$

then converting to Clifford formalism with the use of the $Q^\pm$ operators as in Sect. 3.2.3 along with the S, T operators as in Sect. 5.2 (see Eq. 5.18) and the electromagnetic fields in Eq. 6.18, gives:

$$\begin{cases} TQ^+(f^\star + f^-) - \beta TQ^+ f^+ = 0 \\ \alpha SQ^+(f^\star + f^-) - SQ^+ f^+ = 0 \\ \beta TQ^-(f^\star + f^-) - TQ^- f^+ = 0 \\ SQ^-(f^\star + f^-) - \alpha SQ^- f^+ = 0 \end{cases} \tag{6.21}$$

Using the commutativity of the Q^-, Q^+, S, T operators, then adding the upper and lower pairs of equations leads to:

$$\begin{cases} Q^+\{(\alpha S + T)(f^\star + f^-) - (S + \beta T)f^+\} = 0 \\ Q^-\{(S + \beta T)(f^\star + f^-) - (\alpha S + T)f^+\} = 0 \end{cases} \tag{6.22}$$

Introducing for notational convenience the space-time rescaling operator $R(\nu, \xi)$ which has the properties:

$$\begin{cases} R(\nu,\xi) & = \nu S + \xi T \\ R(\nu,\xi)R(\nu',\xi') & = R(\nu\nu',\xi\xi') \\ R(\nu,\xi)+R(\nu',\xi') & = R(\nu+\nu',\xi+\xi') \end{cases} \tag{6.23}$$

and rewriting Q^+ and Q^- from their original definitions in terms of the reflection operator Q (see Sect. 3.2.1) then gives:

$$\begin{cases} \frac{1}{2}(I+Q)\{R(\alpha,1)(f^\star+f^-)-R(1,\beta)f^+\}=0 \\ \frac{1}{2}(I-Q)\{R(1,\beta)(f^\star+f^-)-R(\alpha,1)f^+\}=0 \end{cases} \tag{6.24}$$

Adding the two equations, utilising the commutativity of the Q and R operators (because R is composed of S and T, and Q is composed of Q^- and Q^-; all of which are commutative) and the addition property of the R operator in Eq. 6.23, gives:

$$\tfrac{1}{2}R(\alpha+1,1+\beta)(f^\star+f^--f^+)+\tfrac{1}{2}R(\alpha-1,1-\beta)Q(f^\star+f^-+f^+)=0 \tag{6.25}$$

Multiplying by the inverse of $R(\alpha+1, 1+\beta)$:

$$\begin{aligned} &\tfrac{1}{2}\left(I+R\left(\tfrac{\alpha-1}{\alpha+1},\tfrac{1-\beta}{1+\beta}\right)Q\right)f^\star+\tfrac{1}{2}\left(I+R\left(\tfrac{\alpha-1}{\alpha+1},\tfrac{1-\beta}{1-\beta}\right)Q\right)f^- \\ &\quad-\tfrac{1}{2}\left(I-R\left(\tfrac{\alpha-1}{\alpha+1},\tfrac{1-\beta}{1+\beta}\right)Q\right)f^+=0 \end{aligned} \tag{6.26}$$

Equation 6.26 plays the role of the Clifford-valued boundary condition for the case of partial transmission and reflection in the same way that the simpler Eqs. 6.6 and 6.13 play that role for the two cases of perfect reflection and transmission. Under the appropriate conditions, both Eqs. 6.6 and 6.13 are recovered from Eq. 6.26.

For perfect transmission the material properties are the same on both sides of the interface ($\alpha=\beta=1$). In this case $\alpha/\beta=\eta_+/\eta_-=1$ so that the coefficient of transmission τ (see Sect. 6.1) takes the value 1, and the coefficient of reflection Γ takes the value zero. No field is reflected ($f^-=0$). The space-time rescaling operator takes the value $R_{\alpha\beta}=R\left(\frac{\alpha-1}{\alpha+1},\frac{1-\beta}{1+\beta}\right)=0$. Substituting into Eq. 6.26 and then applying the operator Q^+ and doubling retrieves Eq. 6.13.

For perfect reflection from a good conductor no magnetic flux can penetrate ($\mu_+=0$) and the effective permittivity $\varepsilon_+=\epsilon_+-i\sigma_+/\omega$ is dominated by the high value of electrical conductivity $\sigma_+\to\infty$. In this case $\alpha=1/\beta=0$ and $\alpha/\beta=\eta_+/\eta_-=0$ so that the coefficient of reflection $\Gamma=-1$, and the coefficient of transmission $\tau=0$. No field is transmitted ($f^+=0$) and the negative sign on Γ indicates that the reflected wave is inverted. The space-time rescaling operator takes the value $R_{\alpha\beta}=-1$. Substituting into Eq. 6.26 retrieves Eq. 6.6 directly.

The expression:

$$g = -\tfrac{1}{2}\left(I + R\left(\tfrac{\alpha-1}{\alpha+1}, \tfrac{1-\beta}{1+\beta}\right) Q\right) f^{\star} \tag{6.27}$$

can be calculated directly from the incident field and taken as data for any particular problem. The boundary condition in Clifford form, Eq. 6.26, is then written more compactly as:

$$\tfrac{1}{2}\left(I + R\left(\tfrac{\alpha-1}{\alpha+1}, \tfrac{1-\beta}{1+\beta}\right) Q\right) f^{-} - \tfrac{1}{2}\left(I - R\left(\tfrac{\alpha-1}{\alpha+1}, \tfrac{1-\beta}{1+\beta}\right) Q\right) f^{+} = g \tag{6.28}$$

The equation can be read to state *"a linear combination of certain components of the reflected and transmitted fields on the left matches certain components of the incident field on the right"*.

6.5 Exercises

Q1. Show that the reflection coefficient Γ can be calculated directly from the contrast factors α, β in the form $\Gamma = \frac{\alpha-\beta}{\alpha+\beta}$.

Q2. Show that the transmission coefficient τ can be calculated directly from the contrast factors α, β in the form $\Gamma = \frac{2\alpha}{\alpha+\beta}$.

Q3. Develop the Clifford-valued boundary condition for the case of perfect transmission using the last two equations from the middle section of Table 6.3.

Q4. By returning to the original definitions of the projection operators Q^{-}, Q^{+} and S, T, show that the first pair commutes with the second pair.

Q5. Verify the behaviour of the operator $R(\nu, \xi) = \nu S + \xi T$ under addition and multiplication as listed in Eq. 6.23.

Q6. By returning to the original definitions of the operators Q and R appearing in Eq. 6.24, show that the first commutes with the second.

Q7. Show that by substituting the conditions $f^{-} = 0$ and $\alpha = \beta = 1$ in the boundary condition 6.26 retrieves the boundary condition 6.13 for the case of perfect transmission.

Q8. Show that by substituting the conditions $f^{+} = 0$, $\alpha = 0$ and $\beta = \infty$ in the boundary condition 6.26 retrieves the boundary condition 6.6 for the case of perfect reflection.

Q9. Verify that $\alpha/\beta = \eta_{+}/\eta_{-} = c_{-}/c_{+}$ where $c_{\pm}$ is the speed of propagation of electromagnetic waves in the two regions $\Omega^{\pm}$.

Chapter 7
Cauchy Integrals

The Cauchy integral for bounded regions in multiple dimensions is created by combining the boundary theorem with the fundamental solution for a point source in an unbounded region. In the case of electromagnetism, the Cauchy integral gives the solution to Maxwell's homogeneous equation:

$$\mathcal{D}_k f = 0 \tag{7.1}$$

for the k-monogenic function $f = \mathcal{F}_k$ in single bounded regions where there are no sources. For cases where there are two or more regions with different material properties, other integral operators known as Hardy projections (cf. [38]) are used instead. The Hardy projections are related to and defined from the Cauchy integral (see Chap. 8).

7.1 Boundary Theorem

The *boundary theorem* is a statement of the fundamental theorem of integral calculus, to the effect that the integral of a function f over a boundary is related to the integral of the derivative Df of the function throughout the region enclosed [2, 30]. For two Clifford-valued functions f and g defined at points $\mathbf{s}$ on the boundary Σ with outward unit normal vector n, and throughout the region Ω^+ enclosed (see Fig. 7.1), the boundary theorem takes the form:

$$\int_\Sigma g(\mathbf{s})n(\mathbf{s})f(\mathbf{s})\, d\sigma(\mathbf{s}) = \int_{\Omega^+} [(gD)(\mathbf{s})f(\mathbf{s}) + g(\mathbf{s})(Df)(\mathbf{s})]\, d\mathbf{s} \tag{7.2}$$

Here $d\mathbf{s}$ represents the elemental measure of m-dimensional space within region Ω^+, and $d\sigma(\mathbf{s})$ represents the elemental measure of the $(m-1)$-dimensional boundary Σ.

A. Seagar, *Application of Geometric Algebra to Electromagnetic Scattering*,
DOI 10.1007/978-981-10-0089-8_7

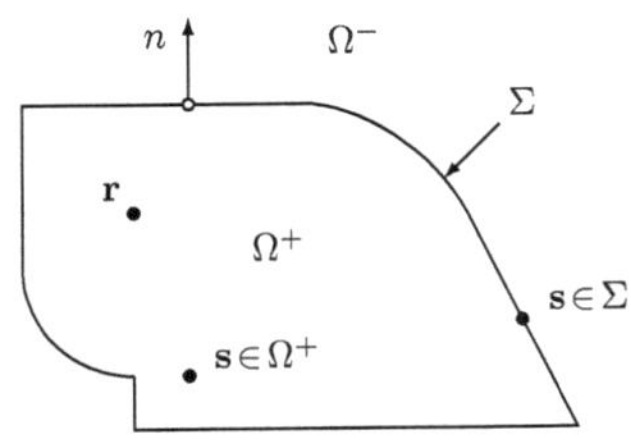

Fig. 7.1 Region Ω^+ enclosed by boundary Σ with normal n. The boundary serves as an interface between the interior Ω^+ and the exterior Ω^-. Point **r** takes any position whereas point **s** takes positions only within and on the boundary

In two dimensions $d\mathbf{s}$ and $d\sigma(\mathbf{s})$ play the role of area and distance respectively. In three dimensions they play the role of volume and area instead. The normal vector n is included in the integrand explicitly so that the elemental measure of boundary $d\sigma$ plays the role of a scalar.

The integrand in brackets on the right hand side takes the form of the Dirac derivative of the product of the two functions, $\mathcal{D}_0(gf)$, expanded by the product rule in Sect. 5.4. Either function f or g can be set to unity to recover simpler forms with one function only, and differentiation by D or multiplication by n from left or right.

7.2 Fundamental Solutions

Solutions for point sources in unbounded regions are called *fundamental* solutions. For any differential operator D the fundamental solution $f(\mathbf{r})$ satisfies the equation:

$$Df(\mathbf{r}) = \delta(\mathbf{r} - 0) \tag{7.3}$$

where δ is the Dirac delta function. From the behaviour of the Dirac delta function the fundamental solution inherits the properties:

$$\begin{cases} Df(\mathbf{r} \neq 0) = 0 \\ \displaystyle\int_\Omega Df(\mathbf{r})\, d\mathbf{r} = 1 \end{cases} \tag{7.4}$$

where $d\mathbf{r}$ is the elemental measure of space within a region Ω covering the origin. Together, these properties are useful in collapsing integrals over regions down to values at single points, as in Sect. 7.3.1.

For the electromagnetic application here, the differential operators D of most interest are the k-Dirac operator $\mathcal{D}_k = \nabla + ke_0$ and the Helmholtz operator $H_k = \nabla^2 + k^2$. The relationship between these two operators $\mathcal{D}_k^2 = -H_k$ (see Sect. 5.3) also entails a relationship between the fundamental solutions of each. In particular, if the fundamental solution of the k-Dirac operator is F_k and the fundamental solution of the Helmholtz operator is B_k then starting with the former:

$$\mathcal{D}_k F_k(\mathbf{r}) = \delta(\mathbf{r} - 0) = H_k B_k(\mathbf{r}) = (-\mathcal{D}_k^2) B_k(\mathbf{r}) = \mathcal{D}_k(-\mathcal{D}_k B_k)(\mathbf{r}) \tag{7.5}$$

gives an expression from which it can be derived by the direct differentiation of the latter:

$$F_k = -\mathcal{D}_k B_k \tag{7.6}$$

The fundamental solution B_k of the Helmholtz operator is called the Bessel potential, and can be written [34] for dimensions $m > 1$:

$$B_k(\mathbf{r}) = \begin{cases} \left(-\frac{1}{2\pi r}\frac{\partial}{\partial r}\right)^{\frac{m-1}{2}} \left(\frac{1}{2ik} e^{ikr}\right) & m \text{ odd} \\ \left(-\frac{1}{2\pi r}\frac{\partial}{\partial r}\right)^{\frac{m-2}{2}} \left(-\frac{i}{4} H_0^{(1)}(kr)\right) & m \text{ even} \end{cases} \tag{7.7}$$

where $H_0^{(1)}(kr)$ is the Hankel function of the first kind of order zero, and $r = |\mathbf{r}|$.

A complementary fundamental solution of the k-Dirac operator can be obtained from the Bessel potential with negative wavenumber k:

$$F_k = -\mathcal{D}_k B_{-k} \tag{7.8}$$

since $H_k = H_{-k}$. The two different solutions, Eqs. 7.6 and 7.8, correspond in the time domain to point sources for which the waves travel in opposite directions; either away from or towards the source. The direction of travel can be determined by reconstructing the full space-time electromagnetic field $F(\mathbf{r}, t)$ from the phasor $F_k(\mathbf{r})$ and the complex signal $e^{i\omega t}$ as in Eq. 5.19.

7.2.1 Three Dimensions

In three dimensions the fundamental solution of the Helmholtz operator is given from Eq. 7.7 with $m = 3$:

$$B_{\pm k}(\mathbf{r}) = -\frac{1}{2\pi r}\frac{\partial}{\partial r}\left(\pm\frac{1}{2ik} e^{\pm ikr}\right) = -\frac{1}{4\pi r} e^{\pm ikr} \tag{7.9}$$

When the complex signal $e^{i\omega t}$ is reintroduced for the complementary form of the solution:

$$B_{-k}(\mathbf{r}) e^{i\omega t} = -\frac{1}{4\pi r} e^{i(\omega t - kr)} \tag{7.10}$$

the phase of the corresponding travelling wave is:

$$\phi = \omega t - kr \tag{7.11}$$

A point of constant phase, for example a crest of the wave, travels at velocity $dr/dt = \omega/k > 0$ away from the source at $r = 0$. In contrast, the wave corresponding to the fundamental solution $B_{+k}(\mathbf{r})$ has phase $\phi = \omega t + kr$ and travels at velocity $dr/dt = -\omega/k < 0$, towards the source.

The fundamental solution for the k-Dirac operator for a waveform travelling away from the source is given by differentiating Eq. 7.9 according to Eq. 7.8. Firstly using both chain and product rules for differentiation until no longer possible:

$$F_k(\mathbf{r}) = -(\nabla + ke_0)(-\tfrac{1}{4\pi r}e^{-ikr}) = \left(-\tfrac{1}{r^2}(\nabla r)e^{-ikr} + \tfrac{1}{r}(-ik)e^{-ikr}\nabla r + ke_0\tfrac{1}{r}e^{-ikr}\right)\tfrac{1}{4\pi} \tag{7.12}$$

and then expanding $\nabla r = \mathbf{r}/|\mathbf{r}| = \mathbf{r}/r = \hat{\mathbf{r}}$ produces:

$$F_k(\mathbf{r}) = a\mathbf{r} + be_0 = \left(-(1 + ikr)\hat{\mathbf{r}} + kre_0\right)\tfrac{e^{-ikr}}{4\pi r^2} = -\left(\hat{\mathbf{r}} + ikr(\hat{\mathbf{r}} + ie_0)\right)\tfrac{e^{-ikr}}{4\pi r^2} \tag{7.13}$$

7.3 Off the Boundary

7.3.1 *Reproducing Formula*

A k-monogenic function f (such as the electromagnetic field $\mathcal{F}_k$) at a point $\mathbf{r}$ within a region Ω^+ enclosed by a boundary Σ can be calculated from the boundary theorem 7.2 provided that the value of the function is known at all points on the boundary and the function g is chosen in a particular way. Firstly, putting:

$$Df = \nabla f = \mathcal{D}_0 f = (\mathcal{D}_k - ke_0)f = -ke_0 f \tag{7.14}$$

into the boundary theorem, gives:

$$\begin{aligned}\int_\Sigma g(\mathbf{s})n(\mathbf{s})f(\mathbf{s})\,d\sigma(\mathbf{s}) &= \int_{\Omega^+}[(g\nabla)(\mathbf{s})f(\mathbf{s}) + g(\mathbf{s})(\nabla f)(\mathbf{s})]\,d\mathbf{s}\\ &= \int_{\Omega^+}(g(\nabla - ke_0))(\mathbf{s})f(\mathbf{s})\,d\mathbf{s}\end{aligned} \tag{7.15}$$

The function $g(\mathbf{s})$ is chosen as:

$$g(\mathbf{s}) = E_k(\mathbf{s} - \mathbf{r}) \tag{7.16}$$

where $\mathbf{r}$ is a fixed point inside Ω^+, and E_k is defined such that its behaves like a fundamental solution under differentiation from the right by the operator $\nabla - ke_0$:

$$(E_k(\nabla - ke_0))(\mathbf{s}) = \delta(\mathbf{s} - 0) \tag{7.17}$$

In that case $(g(\nabla - ke_0))(\mathbf{s}) = \delta(\mathbf{s} - \mathbf{r})$ and the integral over Ω^+ then collapses, leaving:

$$\int_\Sigma E_k(\mathbf{s} - \mathbf{r})n(\mathbf{s})f(\mathbf{s})\,d\sigma(\mathbf{s}) = f(\mathbf{r}) \quad ; \mathbf{r} \in \Omega^+ \tag{7.18}$$

as the value of the function f at the fixed point $\mathbf{r}$.

Equation 7.18 is called the *reproducing formula* [2], because it reproduces the k-monogenic function f within Ω^+ if its value is known on the boundary Σ. As the point $\mathbf{r}$ approaches the boundary the value of f approaches the value on the boundary. The function E_k is called the *Cauchy kernel*. It can be obtained by noting that it is related through the delta function to the fundamental solution F_k:

$$E_k(\nabla - ke_0) = \delta = (\nabla + ke_0)F_k \tag{7.19}$$

The latter is already known as $F_k = a\mathbf{r} + be_0$ with values of a, b as in Eq. 7.13. Expanding the right hand side of Eq. 7.19 and invoking the anticommutative property of Clifford units, gives:

$$\begin{aligned}(\nabla + ke_0)(a\mathbf{r} + be_0) &= \nabla a\mathbf{r} + \nabla be_0 + ke_0a\mathbf{r} + ke_0be_0 \\ &= a\mathbf{r}\nabla - be_0\nabla - a\mathbf{r}ke_0 + be_0ke_0 = (a\mathbf{r} - be_0)(\nabla - ke_0)\end{aligned} \tag{7.20}$$

Replacing a, b with their values from Eq. 7.13 now gives the Cauchy kernel:

$$E_k(\mathbf{r}) = -\left(\hat{\mathbf{r}} + ikr\left(\hat{\mathbf{r}} - ie_0\right)\right)\frac{e^{-ikr}}{4\pi r^2} = -F_k(-\mathbf{r}) \tag{7.21}$$

In cases where the function f is not k-monogenic or the point $\mathbf{r}$ is not within the region Ω^+, the integral over Ω^+ gives different values.

If the function f remains k-monogenic but the point $\mathbf{r}$ is placed in the region Ω^- outside the boundary, the delta function $\delta(\mathbf{s} - \mathbf{r})$ takes the value zero for all $\mathbf{s} \in \Omega^+$, and the integral over Ω^+ collapses to zero:

$$\int_\Sigma E_k(\mathbf{s} - \mathbf{r})n(\mathbf{s})f(\mathbf{s})\,d\sigma(\mathbf{s}) = 0 \quad ; \mathbf{r} \in \Omega^- \tag{7.22}$$

Alternatively, if the point $\mathbf{r}$ is placed exactly on a smooth part of the boundary (where the tangent plane is unambiguous) the delta function behaves as if half inside Ω^+ and half inside Ω^-. In this case the integral collapses to one-half of the value within Ω^+ immediately adjacent to the boundary:

$$\int_{\Sigma} E_k(\mathbf{s}-\mathbf{r})n(\mathbf{s})f(\mathbf{s})\,d\sigma(\mathbf{s}) = \tfrac{1}{2}f(\mathbf{r}); \quad \mathbf{r} \in \Sigma \tag{7.23}$$

If the function f is not k-monogenic (i.e. not consistent with an electromagnetic field) the values of f produced by the integral over Ω^+ are k-monogenic but do not approach the (non-monogenic) values on the boundary as $\mathbf{r}$ approaches Σ. This behaviour is useful when the function f on the boundary is a field which has been measured experimentally, and may be contaminated with some amount of largely random error. Any error which is inconsistent with an electromagnetic field is effectively filtered out and rejected by the reproducing formula. For example, if the near field radiation pattern of an antenna is measured and the finite precision of the measurements introduces error, the operation of calculating either the closely adjacent near field or the far field radiation pattern using Eq. 7.18, as in Sect. 7.5, serves to attenuate the error.

7.3.2 Cauchy Extension

Figure 7.2 shows electromagnetic fields generated by two point sources, δ^+ within Ω^- and δ^- within Ω^+.

The reproducing formula regenerates the monogenic field f^+ within the region Ω^+, extending away from the source δ^+, from the trace of the field on the boundary. However, the reproducing formula returns a value of zero in the region Ω^- where the field extending back towards the source is non-monogenic:

$$\int_{\Sigma} E_k(\mathbf{s}-\mathbf{r})n^+(\mathbf{s})f^+(\mathbf{s})\,d\sigma(\mathbf{s}) = \begin{cases} f^+(\mathbf{r}) & ; \mathbf{r} \in \Omega^+ \\ \frac{1}{2}f^+(\mathbf{r}) & ; \mathbf{r} \in \Sigma \\ 0 & ; \mathbf{r} \in \Omega^- \end{cases} \tag{7.24}$$

Similarly, with a reversed value of the normal $n^- = -n^+$, the reproducing formula regenerates the monogenic field f^- within the region Ω^-, extending away from the source δ^-, from its trace on the boundary:

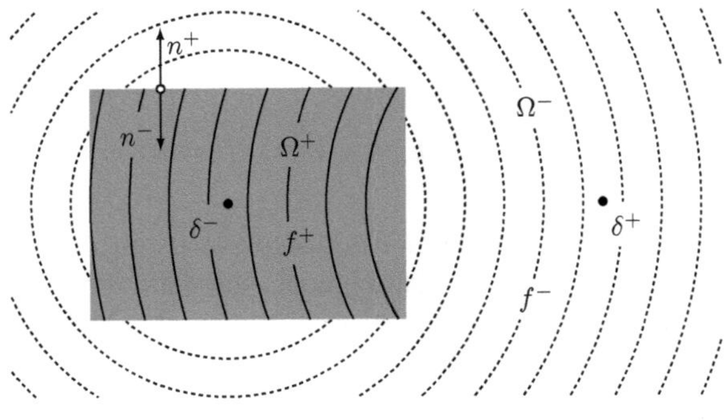

Fig. 7.2 Fields generated by reproducing formula from boundary traces of fields due to internal and external sources δ^- and δ^+ respectively

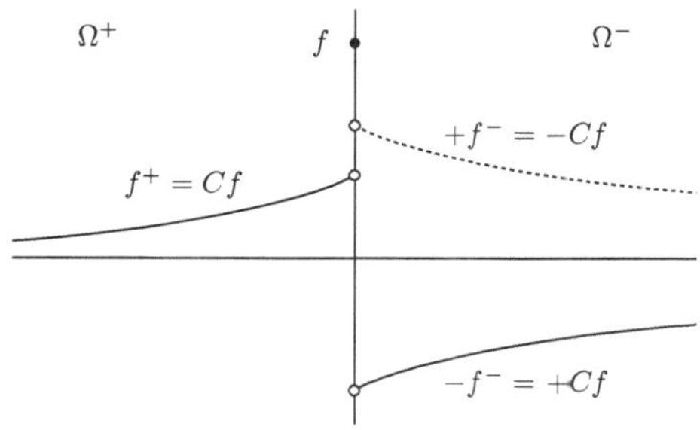

Fig. 7.3 Fields generated by Cauchy extension from function f on boundary

$$\int_\Sigma E_k(\mathbf{s}-\mathbf{r})n^-(\mathbf{s})f^-(\mathbf{s})\,d\sigma(\mathbf{s}) = \begin{cases} 0 & ;\ \mathbf{r}\in\Omega^+ \\ \frac{1}{2}f^-(\mathbf{r}) & ;\ \mathbf{r}\in\Sigma \\ f^-(\mathbf{r}) & ;\ \mathbf{r}\in\Omega^- \end{cases} \tag{7.25}$$

A single formulation is created by subtracting Eq. 7.25 from Eq. 7.24:

$$(Cf)(\mathbf{r}) = \int_\Sigma E_k(\mathbf{s}-\mathbf{r})n(\mathbf{s})f(\mathbf{s})\,d\sigma(\mathbf{s}) = \begin{cases} f^+(\mathbf{r}) & ;\ \mathbf{r}\in\Omega^+ \\ \frac{1}{2}[f^+(\mathbf{r}) - f^-(\mathbf{r})] & ;\ \mathbf{r}\in\Sigma \\ -f^-(\mathbf{r}) & ;\ \mathbf{r}\in\Omega^- \end{cases} \tag{7.26}$$

where $n = n^+ = -n^-$ and $f = f^+ + f^-$. The operator C is called the *Cauchy extension* [2], and the function f on the interface where the two regions share the common boundary Σ, is called the *joint interface function*.

The Cauchy extension, extends the interface function f defined on the common boundary into the regions on both sides (see Fig. 7.3). Taking the direction of the normal $n = n^+$ to define Ω^+ as the inside, the Cauchy extension extends the function f inward as f^+, and outward as minus f^-.

The extensions do not approach the value f on the boundary as the point $\mathbf{r}$ approaches Σ. The fields f^+ and f^- are usually discontinuous across the boundary, both in value and derivative. It is possible to have continuity if neither Ω^+ or Ω^- extends to infinity but is otherwise impossible unless both f^+ and f^- are identically zero.

7.4 On the Boundary

7.4.1 Cauchy Integral

For the special case of points $\mathbf{r}\in\Sigma$ on the boundary, the Cauchy extension is used to construct an additional integral operator:

$$(C_\Sigma f)(\mathbf{r}) = 2\,\text{p.v.}\int_\Sigma E_k(\mathbf{s}-\mathbf{r})n(\mathbf{s})f(\mathbf{s})\,d\sigma(\mathbf{s}) = f^+(\mathbf{r}) - f^-(\mathbf{r});\qquad \mathbf{r}\in\Sigma \tag{7.27}$$

where "p.v." indicates that the Cauchy principal value of the integral is taken to properly account for the singularity in the Cauchy kernel E_k at $\mathbf{s} = \mathbf{r}$. The operator C_Σ is called the *principal value Cauchy integral* [2, 38].

For sources δ^+ and δ^- the principal value Cauchy integral operator C_Σ gives:

$$C_\Sigma(f^+ + f^-) = f^+ - f^- \tag{7.28}$$

For sources δ^+ and $-\delta^-$ the second field is inverted so that:

$$C_\Sigma(f^+ - f^-) = f^+ + f^- \tag{7.29}$$

Together Eqs. 7.28 and 7.29 show that applying the Cauchy integral twice to the function at a point on the boundary returns the original value:

$$C_\Sigma^2 = I \tag{7.30}$$

As a consequence, this particular form of the Cauchy integral has the important property of behaving like a reflection operator (see Sect. 3.1). Although similar, lacking the extra factor of two the Cauchy extension when evaluated on the boundary is not a reflection operator.

The reflection property of the principal value Cauchy integral operator is exploited to define the Hardy projection operators from the Cauchy integral in Chap. 8, using the same method as in Sect. 3.1.

7.5 Example Application

The integral operators described in this chapter and in Chap. 8 can be used in different ways to solve different kinds of electromagnetic problems. The simplest problem of practical interest involves the migration of a field measured on one surface enclosing a source, to another surface enclosing the first. This situation is encountered when the radiation pattern is measured for convenience at a particular distance from an antenna and the objective is to calculate from that, the pattern which would be measured on another surface some distance further away from the antenna. In this case the reproducing formula is applied directly by numerical evaluation of the integral in Eq. 7.18. As long as the point $\mathbf{r}$ at which the field is evaluated is not too close to the surface, it is not necessary to adopt a particularly sophisticated numerical method.

7.5.1 Test Case

Figure 7.4 shows a dipole source δ^+ located at the centre of a cube with sides aligned to a Cartesian coordinate system x, y, z. The source generates sinusoidal electric

Fig. 7.4 Dipole source δ^+ enclosed by cubic surface. The surface is subdivided into elements Σ_j

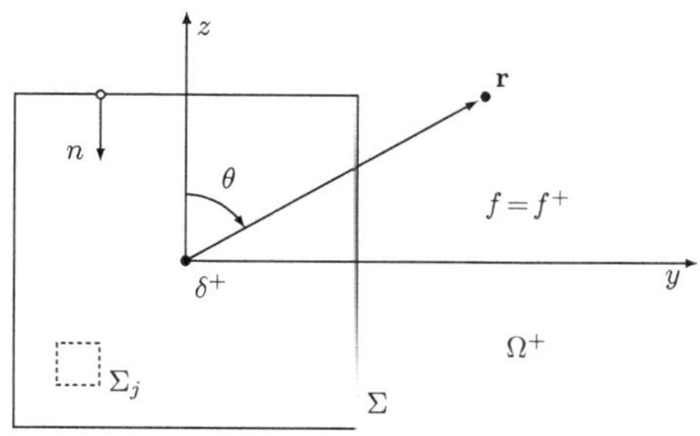

and magnetic fields written in vector notation **E**, **H** in spherical coordinates r, θ, ϕ as [29]:

$$\begin{cases} \mathbf{E}(r,\theta) = \frac{Ih}{4\pi r^2}\eta\left\{2\left(1+\frac{1}{ikr}\right)e^{-ikr}(\cos\theta)\,\hat{\mathbf{r}} + \left(1+ikr+\frac{1}{ikr}\right)e^{-ikr}(\sin\theta)\,\hat{\boldsymbol{\theta}}\right\} \\ \mathbf{H}(r,\theta) = \frac{Ih}{4\pi r^2}(1-ikr)e^{-ikr}(\sin\theta)\,\hat{\boldsymbol{\phi}} \end{cases} \tag{7.31}$$

The cube is completely transparent with material properties of electric permittivity, permeability and conductivity the same on the inside and the outside. The field transmitted outside the cube is a completely continuous extension of the field from the dipole incident on the inner surface of the cube. There is no reflected field.

The surface of the cube is identified by a boundary Σ with outward normal n. Note that the outward normal to the boundary is directed towards the interior of the cube as shown in Fig. 7.4 so that the reproducing formula gives the value $f = f^+ \neq 0$ at the point **r** in the region Ω^+ some distance further from the source than is the boundary Σ.

This test case has been used as a standard by Chantaveerod [39, 40], with comparison to measured data [41–43].

7.5.2 Numerical Method

The field $f(\mathbf{r})$ can be calculated using the reproducing formula by subdividing the surface into individual elements Σ_j and summing the fields $f_j(\mathbf{r})$ from each element:

$$\begin{cases} f_j(\mathbf{r}) = \int_{\Sigma_j} E_k(\mathbf{s}-\mathbf{r})n(\mathbf{s})f(\mathbf{s})\,d\sigma(\mathbf{s}) \\ f(\mathbf{r}) = \sum_j f_j(\mathbf{r}) \end{cases} \tag{7.32}$$

If the field is evaluated at points **r** sufficiently far from each element compared to their size, and if in addition each element is made sufficiently small compared to the wavelength λ of the field, then the value of the Cauchy kernel and the field each take approximately uniform values $E_k(\mathbf{s}_j - \mathbf{r})$ and $f(\mathbf{s}_j)$ respectively over the whole element, for some fixed point $\mathbf{s}_j$.

In this case the whole integrand takes a constant value, and the field produced by a single element can be written:

$$f_j(\mathbf{r}) \approx E_k(\mathbf{s}_j - \mathbf{r})n(\mathbf{s}_j)f(\mathbf{s}_j)A_j \tag{7.33}$$

where A_j is the area of the element. The total field is then calculated using the sum in Eq. 7.32.

The approximation is only good when the assumptions are honoured.

7.5.3 Accuracy of Results

Figure 7.5 shows the RMS (root mean square) value $\sqrt{\mu|\mathbf{H}|^2 + \epsilon|\mathbf{E}|^2}$ of the field calculated for a dipole standing in free space ($\eta = \eta_0 = \sqrt{\mu_0/\epsilon_0}$ ohms) with strength $Ih = 1$ Ampère-metres and wavenumber $k = 2\pi/\lambda = 1$ per metre. The figure also shows the error when taking values of $f(\mathbf{s}_j)$ given by Eq. 7.31 at the centre of each of $N = 6S^2$ elements on the surface of a cube one metre along each edge, and calculating the total field by the summation in Eq. 7.32. The field is calculated with sampling S between 10 and 500 elements per edge (i.e. for N between 600 and 1,500,000 in total), and at points along the x-axis from 1 metre to 1000 metres away from the centre of the cube.

Both absolute field strength and error decrease in inverse proportion to distance. From $r = 10$ to 100 m both reduce by a factor of 10. As a consequence the relative

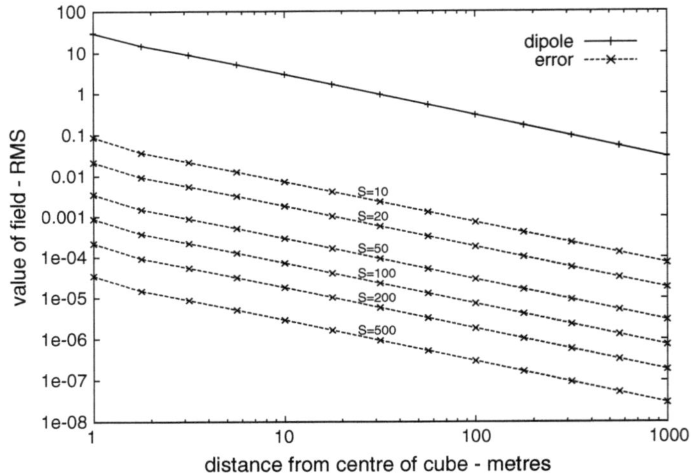

Fig. 7.5 Error in approximate field calculation of Cauchy extension

error remains the same. However, relative error does decrease as the number of elements increases, in inverse proportion. For example, from $N = 6S^2 = 600$ to 60,000 elements the error decreases by a factor of 100.

If higher accuracy is required a more sophisticated numerical method than described here should be used.

7.5.4 Practical Issues

The choice of programming language for generating the results shown in Sect. 7.5.3 was C language. The REDUCE algorithms described in Sect. 2.4.3 and in appendix A are slow, because they are executed by an interpreter rather than being compiled and executed in machine code. Furthermore, the REDUCE algorithms are overly general – they operate for both symbolic and numeric entities and for combinations of Clifford numbers of full generality in any number of dimensions. That is appropriate for demonstrating and verifying algebraic results, but not for numerical calculations.

For calculations in electromagnetism all that is required are numerical operations in four dimensions. Furthermore, the full generality of Clifford multiplication is not required. For the integrand of the integrals in Sects. 7.3 and 7.4, it is only necessary to calculate the product of two bivectors and one vector. Many of the components in these three terms are zero *a priori*, and it is wasteful to carry those parts of the overall Clifford multiplication.

It is much more efficient to construct special purpose four-dimensional hard-wired code which deals only with the multiplication of the non-zero components in these very particular three terms. No recursion need be used. Such code can be modelled on a straightforward one line multiplication of complex numbers to give each of the two components of that product explicitly in terms of all other components. The behaviour of the code could be expressed loosely in words as *"multiply all of these components with all of those components and add them up with the correct sign and put the result in that component"*. Components that are known to be non-zero are not used in the multiplications. However, results that are known to be zero (there is one such grade 0 and grade 4 value) may be calculated as partial checks on the internal consistency of the algorithm.

7.6 Exercises

Q1. Check that $Df = 0$ for $D = H_k$ and $f = B_{-k}$ in two dimensions.

Q2. Check that $\int Df \, d\mathbf{x} = 1$ for $D = H_k$ and $f = B_{-k}$ in two dimensions.

Q3. Check that $Df = 0$ for $D = H_k$ and $f = B_{-k}$ in three dimensions.

Q4. Check that $Df = 0$ for $D = D_k$ and $f = F_k$ in three dimensions.

Q5. By introducing $H_k B_{-k} = H_k B_k$ into Eq. 7.5 show that the fundamental solution of the k-Dirac operator can be calculated using the Bessel potential with negative wavenumber.

Q6. Show $\nabla r = \mathbf{r}/r$ by putting $r = \sqrt{r_x^2 + r_y^2 + r_z^2}$ and expanding in Cartesian coordinates.

Q7. Verify that $\nabla \mathbf{r} = \mathbf{r}\nabla$ by putting $\mathbf{r} = r_x e_1 + r_y e_2 + r_z e_3$ and expanding in Cartesian coordinates.

Q8. Implement the numerical method described in Sect. 7.5.2 for the test case described in Sect. 7.5.1 and reproduce the results shown in Fig. 7.5.

Q9. Modify your implementation in question 8 to add random errors to the field on the surface of the cube. Investigate how well the reproducing formula eliminates the error when calculating the field away from the cube.

Q10. Develop a numerical method for the test case described in Sect. 7.5.1 to reproduce the field accurately in the region from zero to one metre from the surface of the cube.

Chapter 8
Hardy Projections

Two projection operators called *Hardy projections* are constructed from the principal value Cauchy integral C_Σ (*cf.* [2, 38]):

$$\begin{cases} P^+ = \frac{1}{2}(I + C_\Sigma) \\ P^- = \frac{1}{2}(I - C_\Sigma) \end{cases} \tag{8.1}$$

The Hardy projections are the values approached by the Cauchy extension for a point which approaches the boundary from either side. They serve the purpose of splitting the joint interface function $f = f^+ + f^-$ into transmitted and reflected components, f^+ and f^- respectively:

$$\begin{cases} P^+ f = \frac{1}{2}((f^+ + f^-) + (f^+ - f^-)) = f^+ \\ P^- f = \frac{1}{2}((f^+ + f^-) - (f^+ - f^-)) = f^- \end{cases} \tag{8.2}$$

which, as shown in Fig. 8.1, are consistent with the traces on the boundary of electromagnetic fields in the two regions on either side.

The Hardy projections in Eq. 8.2 serve to enforce Maxwell's equations, which are buried inside many layers deep; beneath the Cauchy integral (Sect. 7.4.1) then the Cauchy extension (Sect. 7.3.2) then the reproducing formula (Sect. 7.3.1) and then the boundary theorem (Sect. 7.1) as invoked with the electromagnetic field and k-Dirac operator (both from Maxwell's equations in Sect. 5.3) and the Cauchy kernel (Sect. 7.3.1, from the fundamental solution of the k-Dirac operator in three dimensions in Sect. 7.2.1).

Although there are two Hardy projections in Eq. 8.2, they are not independent because $P^+ + P^- = I$ and either can be derived from the other. Which is of more immediate use depends on the nature of the scattering and the boundary conditions (as in Sect. 6.4) which result. The cases of perfect reflection and perfect transmission are straightforward because the material properties are the same everywhere the field is non-zero. In the case of partial transmission and reflection non-zero fields occupy two regions with different values of material properties μ, ϵ. The wavenumber

A. Seagar, *Application of Geometric Algebra to Electromagnetic Scattering*,
DOI 10.1007/978-981-10-0089-8_8

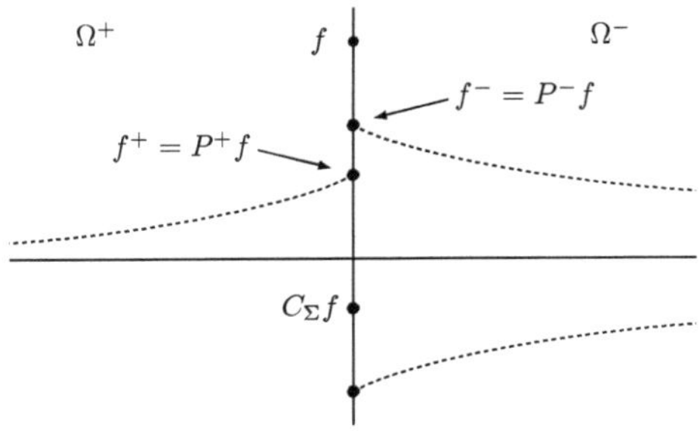

Fig. 8.1 Fields generated by Hardy projection from function f on boundary

$k = 2\pi/\lambda = \omega/c = \omega\sqrt{\mu\epsilon}$ then takes different values on either side of the interface where the two regions share a common boundary. It is necessary to construct Hardy projections for each side of the interface from two different Cauchy integrals, one for each of the different materials.

8.1 Perfect Reflection

In the case of perfect reflection there is no field transmitted through the interface, so that f^+ plays no role in the solution. Recalling for any projection operator that $P^2 = P$ (see Sect. 3.1) and applying the second Hardy projection to the reflected component of the field in Eq. 8.2 leads to:

$$P^- f^- = f^- \tag{8.3}$$

with f^+ eliminated. Equation 8.3 is the form in which Maxwell's equations are honoured for perfect reflection. The solution based on this form is described in Sect. 9.1. It seeks to find the reflected field f^-.

8.2 Perfect Transmission

In the case of perfect transmission there is no field reflected from the interface, so that f^- plays no role in the solution. Applying the first Hardy projection to the transmitted component of the field in Eq. 8.2 leads to:

$$P^+ f^+ = f^+ \tag{8.4}$$

with f^- eliminated. Equation 8.4 is the form in which Maxwell's equations are honoured for perfect transmission. The solution based on this form is described in Sect. 9.2. It seeks to find the transmitted field f^+.

8.3 Partial Transmission and Reflection

In the case of partial transmission and reflection, fields are both transmitted and reflected so that f^+ and f^- are both non-zero. With two regions having different values of material properties, twice as many reflection and projection operators are involved.

Firstly Cauchy extensions C^+ and C^- for are required for regions Ω^+ and Ω^- respectively. These are the same as in Sect. 7.3.2 excepting that the value of $k = k_+ = \omega\sqrt{\mu_+\epsilon_+}$ or $k = k_- = \omega\sqrt{\mu_-\epsilon_-}$ is different for each. Although these extensions do give values of functions on both sides of the interface, only the ones in Fig. 8.2 on the side with the same label, plus or minus, play a direct role in the solution.

From the two Cauchy extensions, two corresponding Cauchy integrals C_Σ^+ and C_Σ^- are constructed, as in Sect. 7.4.1, and then four Hardy projections as in Eq. 8.1:

$$\begin{cases} P^{++} = \frac{1}{2}(I + C_\Sigma^+) \\ P^{+-} = \frac{1}{2}(I - C_\Sigma^+) \end{cases} \quad \text{and} \quad \begin{cases} P^{-+} = \frac{1}{2}(I + C_\Sigma^-) \\ P^{--} = \frac{1}{2}(I - C_\Sigma^-) \end{cases} \tag{8.5}$$

Two of these play an essential role: P^{++} as the boundary limit of C^+ from within region Ω^+ with material properties μ_+, ϵ_+ and wavenumber k_+, and P^{--} as the boundary limit of C^- from within region Ω^- with material properties μ_-, ϵ_- and wavenumber k_-. The other two projections $P^{\pm\mp}$ are not essential for the solution.

For this case of partial transmission and reflection the equations which serve as the forms which honour Maxwell's equations are the new, region specific Hardy projections taken directly as in Eq. 8.2 without further simplification:

$$\begin{cases} P^{++}f = f^+ \\ P^{--}f = f^- \end{cases} \tag{8.6}$$

Here the two projections are different than before, and the unknown joint interface function f which splits into the transmitted and reflected fields under the operation of the projections is no longer simply the sum of the two fields, $f \neq f^+ + f^-$. This is because these two particular Hardy projections are not a complementary pair.

Rather than solving for the joint interface function f as a first step and then calculating the transmitted and reflected fields $f^\pm$ from that using Eq. 8.6, the approach

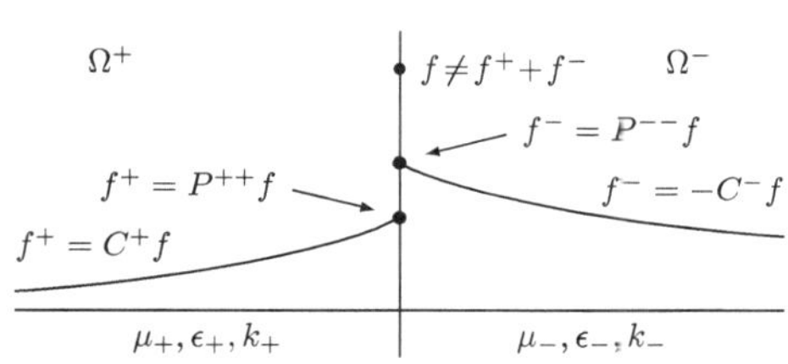

Fig. 8.2 Cauchy extensions and Hardy projections for partial transmission and reflection

described in Sect. 9.3 seeks to find the two fields $f^{\pm}$ directly and simultaneously. Once these are determined on the interface, the fields away from the interface can be recovered from them by application of the two Cauchy extensions $C^{\pm}$, and the solution is effectively complete.

From the fields $f^{\pm}$ it is also possible to calculate the joint interface function f, although that takes about as much work as does finding $f^{\pm}$ in the first instance, with no real benefit. Whereas f serves a mathematical role in the solution it corresponds to no physical field. See Sect. 9.3.4 for details.

8.4 Exercises

Q1. Verify equation 8.3 by applying the Hardy projection P^{-} to the reflected field f^{-} in Eq. 8.2.

Q2. Verify equation 8.4 by applying the Hardy projection P^{+} to the transmitted field f^{+} in Eq. 8.2.

Q3. Show that the two Hardy projections P^{++} and P^{--} are not complementary.

Chapter 9
Construction of Solutions

9.1 Perfect Reflection

The solution for the case of perfect reflection is governed by the boundary conditions in Eq. 6.9 and the Hardy projection of Eq. 8.3:

$$\begin{cases} Q^- f = g \\ P^- f = f \end{cases} \tag{9.1}$$

where $f = f^-$ is the reflected field.

Figure 9.1 shows the solution f as the sum of two components $P^- g$ and $P^- Q^+ f$ along the OP' axis in Banach space (see Sect. 3.4.2). The two components are obtained by splitting f first into $Q^+ f$ and $Q^- f = g$, and then projecting both of these onto the OP' axis:

$$f = P^- g + P^- Q^+ f \tag{9.2}$$

The solution:

$$f = (I - P^- Q^+)^{-1} P^- g \tag{9.3}$$

is expanded as an infinite binomial series:

$$\nu = (1 - \xi)^{-1} = \sum_{j=0}^{\infty} \xi^j \tag{9.4}$$

where $\xi = P^- Q^+$ and 1 stands for the identity operation. The resulting expression:

$$f = \left[\sum_{j=0}^{\infty} (P^- Q^+)^j \right] P^- g \tag{9.5}$$

A. Seagar, *Application of Geometric Algebra to Electromagnetic Scattering*,
DOI 10.1007/978-981-10-0089-8_9

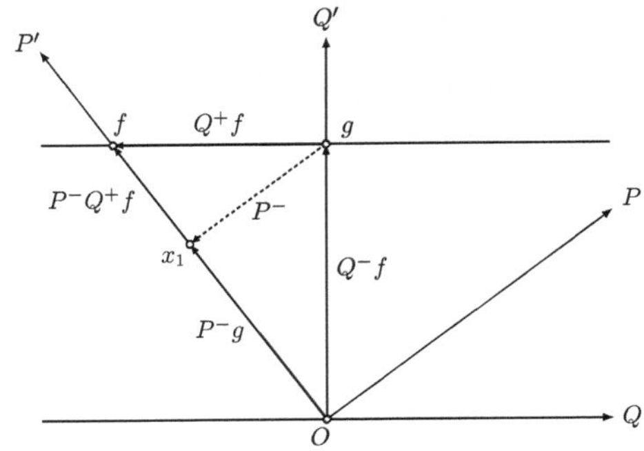

Fig. 9.1 Solution $f = f^-$ as sum of two components along OP' axis

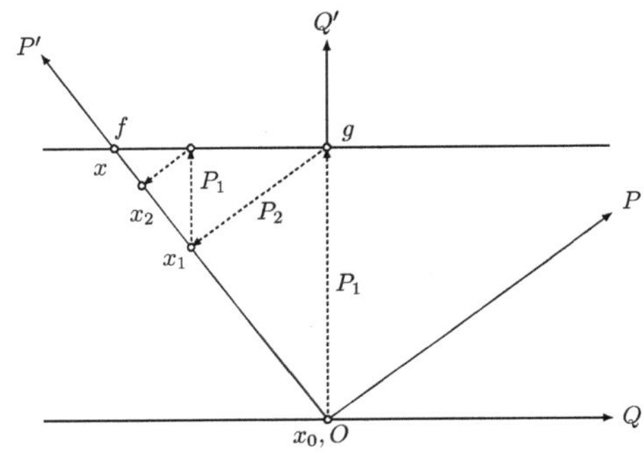

Fig. 9.2 Solution $f = f^-$ as sum of geometric series

is the same as the geometric solution, Eq. 3.34, with $x = f$, $x_0 = 0$, $x_1 = P^-g$ and $\nu = \sum_{j=0}^{\infty}(P^-Q^+)^j$, and with projections $P_1x = Q^+x + g$ and $P_2x = P^-x$ for any x. In terms of the geometric solution, Fig. 9.1 is relabelled as in Fig. 9.2.

An equivalent solution is obtained by casting the series $\sum_{j=0}^{\infty}(P^-Q^+)^j$ into the form:

$$\nu = [1 + \xi + \xi^2 + \xi^3 + \cdots] = [1 \cdots + \xi(1 + \xi(1 + \xi))] \tag{9.6}$$

Horner's nested expansion [44–46] on the right hand side of Eq. 9.6 involves fewer operations and is more numerically stable than the series on the left [47]. This approach has been successfully implemented for electromagnetic scattering by Chantaveerod [40, 48].

Another approach is to use the method of alternating projections onto convex sets [49]:

$$f = \prod_{j=0}^{\infty}(P_2P_1)^j x_0 \tag{9.7}$$

Here the staring point $x_0 = 0$ is chosen, although other starting points can also be used. There are two convex sets, in this case two straight lines. The space is Banach and there is no scalar product, no angles, and the projections onto the sets/lines cannot therefore be orthogonal to them. The points on the lines themselves are multi-dimensional; their dimension being the number of independent variables taken to represent the electromagnetic field over the surface of the scatterer.

9.2 Perfect Transmission

The solution for the case of perfect transmission is governed by the boundary conditions in Eq. 6.16, and the Hardy projection of Eq. 8.4:

$$\begin{cases} Q^+ f = h \\ P^+ f = f \end{cases} \tag{9.8}$$

where $f = f^+$ is the transmitted field.

Figure 9.3 shows the solution f as the sum of two components P^+h and P^+Q^-f along the OP axis in Banach space (see Sect. 3.4.2). The two components are obtained by splitting f first into Q^-f and $Q^+f = h$, and then projecting both of these onto the OP axis:

$$f = P^+h + P^+Q^-f \tag{9.9}$$

The solution:

$$f = (I - P^+Q^-)^{-1}P^+h \tag{9.10}$$

is expanded as an infinite binomial series (see Eq. 9.4). The resulting expression:

$$f = \left[\sum_{j=0}^{\infty}(P^+Q^-)^j\right] P^+h \tag{9.11}$$

is the same as the geometric solution, Eq. 3.34, with $x = f$, $x_0 = 0$, $x_1 = P^+h$ and $\nu = \sum_{j=0}^{\infty}(P^+Q^-)^j$, and with projections $P_1x = Q^-x + h$ and $P_2x = P^+x$ for any x. In terms of the geometric solution, Fig. 9.3 is relabelled as in Fig. 9.4.

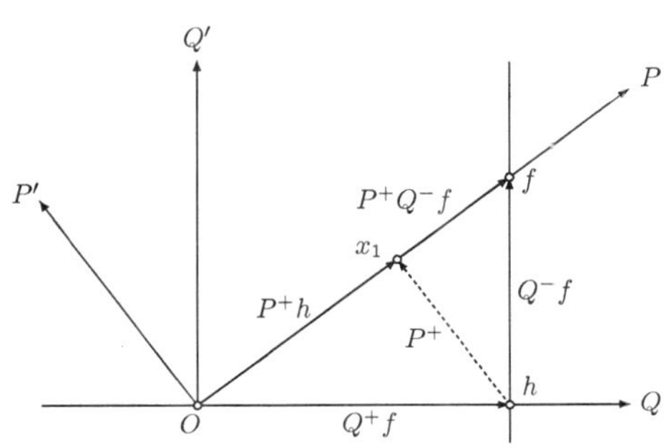

Fig. 9.3 Solution $f = f^+$ as sum of two components along OP axis

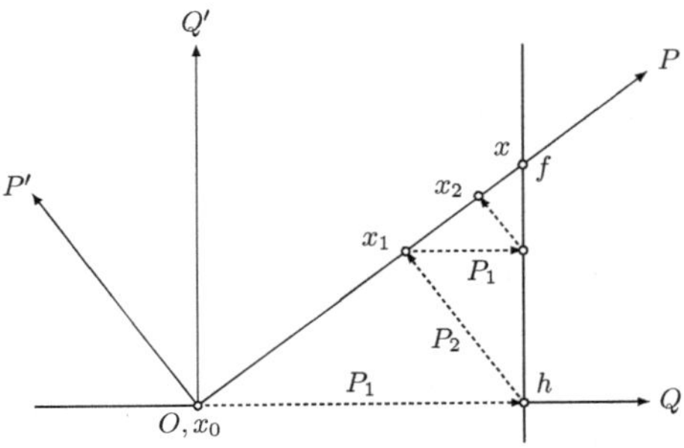

Fig. 9.4 Solution $f = f^+$ as sum of geometric series

An equivalent solution is obtained by casting the series $\sum_{j=0}^{\infty}(P^+Q^-)^j$ into the form:

$$\nu = [1 + \xi + \xi^2 + \xi^3 + \cdots] = [1 \cdots + \xi(1 + \xi(1 + \xi))] \tag{9.12}$$

with $\xi = P^+Q^-$. Horner's nested expansion [44, 46] on the right hand side of Eq. 9.12 involves fewer operations and is more numerically stable than the series on the left [47].

As with the perfect conductor, the method of alternating projections onto convex sets [49] can also be used:

$$f = \prod_{j=0}^{\infty}(P_2 P_1)^j x_0 \tag{9.13}$$

Here the staring point $x_0 = 0$ is chosen, although other starting points can also be used. There are two convex sets, in this case two straight lines. The space is Banach and there is no scalar product, no angles, and the projections onto the sets/lines cannot therefore be orthogonal to them. The points on the lines themselves are multi-dimensional; their dimension being the number of independent variables taken to represent the electromagnetic field over the surface of the scatterer.

9.3 Partial Transmission and Reflection

The solution for the case of partial transmission and reflection is governed by the boundary conditions in Eq. 6.28 and by the Hardy projections in Eq. 8.6:

$$\begin{cases} \frac{1}{2}(I + R_{\alpha\beta}Q)f^- - \frac{1}{2}(I - R_{\alpha\beta}Q)f^+ = g \\ P^{++}f = f^+ \\ P^{--}f = f^- \end{cases} \tag{9.14}$$

where f is the joint interface function, f^+ is the transmitted field, f^- is the reflected field, and $R_{\alpha\beta} = R\left(\frac{\alpha-1}{\alpha+1}, \frac{1-\beta}{1+\beta}\right)$ is the space-time rescaling operator.

The two Hardy projections $P^{\pm\pm}$ are not complimentary so that the two axes they define do not serve as a Cartesian coordinate system. The terms in parentheses on the first line of Eq. 9.14 are not projection operators because $R_{\alpha\beta} \neq \pm I$ for most values of the contrast factors α and β. The space-time rescaling operator $R_{\alpha\beta}$ is neither a projection nor reflection operator, but is linear, so that:

$$\begin{cases} R_{\alpha\beta}(\xi x) = \xi R_{\alpha\beta}(x) \\ R_{\alpha\beta}(x+y) = R_{\alpha\beta}(x) + R_{\alpha\beta}(y) \end{cases} \tag{9.15}$$

As a notational convenience the product of the space-time rescaling operator $R_{\alpha\beta}$ with the boundary reflection operator Q is simply written RQ.

9.3.1 Construction

The geometric constructions of Sects. 9.1 and 9.2 for the special cases of perfect reflection and perfect transmission are generalised according to Eq. 9.14 in the form of Fig. 9.5.

The axes OQ and OQ' serve as a Cartesian coordinate system, whereas the axes $OP^{\oplus}$ and $OP^{\ominus}$ do not. The lines $I^{\oplus}_{RQ}$ and $I^{\ominus}_{RQ}$ are the images of the axes $OP^{\ominus}$ and $OP^{\oplus}$ under the mappings $\frac{1}{2}(I \pm RQ)$. The images are indeed lines because the mappings are linear, but do not serve as axes because the mappings are not projections.

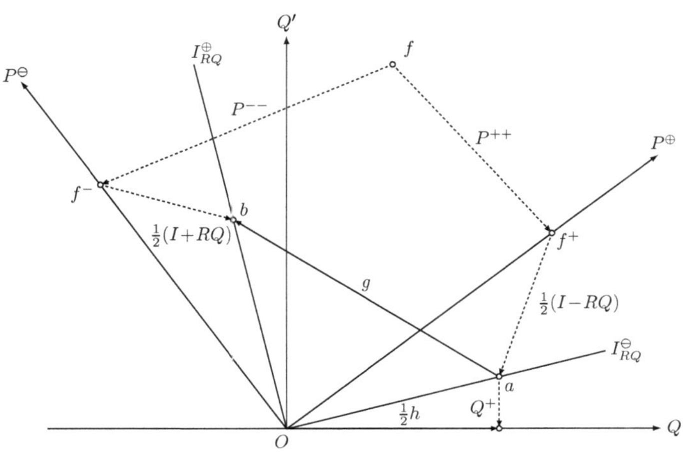

Fig. 9.5 Geometric construction for general solution

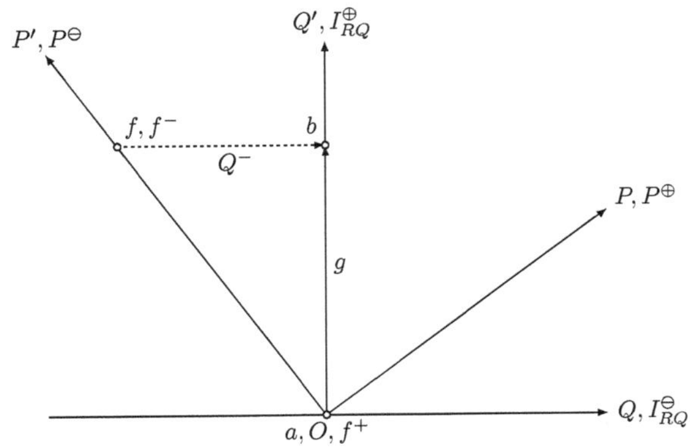

Fig. 9.6 General solution reduced with $\Gamma = -1$ to case of perfect reflection

The general solution of Fig. 9.5 reduces to the particular solutions of Sects. 9.1 and 9.2 as the coefficient of reflection $-1 \leq \Gamma \leq 0$ approaches -1 and 0.

As Γ approaches -1, the axes $OP^{\oplus}$ and $OP^{\ominus}$ migrate onto the coordinate axes OP and OP'. The transmitted field f^+ retracts towards the origin along the $OP^{\oplus}$ axis taking f towards the $OP^{\ominus}$ axis, and merging with it on the OP' axis. One end b of the boundary data vector g migrates with the $I^{\oplus}_{RQ}$ line onto the OQ' axis, while the other end a is driven to the origin along the $I^{\ominus}_{RQ}$ line by the retraction of f^+ to the origin. The $I^{\ominus}_{RQ}$ line migrates onto the OQ axis.

The result in Fig. 9.6 is identical to the special case in Fig. 9.1 with the exception that some of the vectors, points and axes are overloaded with a multiplicity of names and interpretations.

As Γ approaches 0, the coordinate axes $OP^{\oplus}$ and $OP^{\ominus}$ again migrate onto the coordinate axes OP and OP'. The reflected field f^- retracts towards the origin along the $OP^{\ominus}$ axis taking f towards the $OP^{\oplus}$ axis, and merging with it on the OP axis. One end a of the boundary data vector g migrates with the $I^{\ominus}_{RQ}$ line onto the OP axis, while the other end b is driven to the origin along the $I^{\oplus}_{RQ}$ line by the retraction of f^- to the origin. The $I^{\oplus}_{RQ}$ line migrates onto the OP' axis.

The result in Fig. 9.7 is identical to the special case in Fig. 9.3, again with the exception that some of the vectors, points and axes are overloaded with a multiplicity of names and interpretations.

The general case of Fig. 9.5 covers the continuum of solutions between perfect reflection, with coefficient of reflection $\Gamma = -1$, and perfect transmission, with reflection coefficient $\Gamma = 0$.

9.3.2 Solution

The solution in the general case is based on the method of alternating projections onto convex sets. In this case it is the two Hardy projections P^{++} and P^{--} which

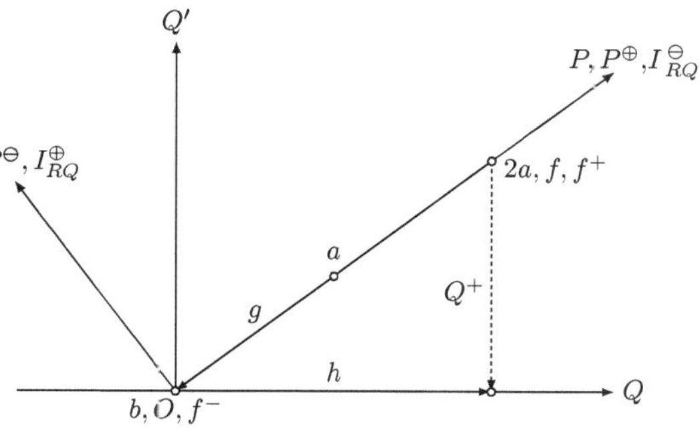

Fig. 9.7 General solution reduced with $\Gamma = 0$ to case of perfect transmission

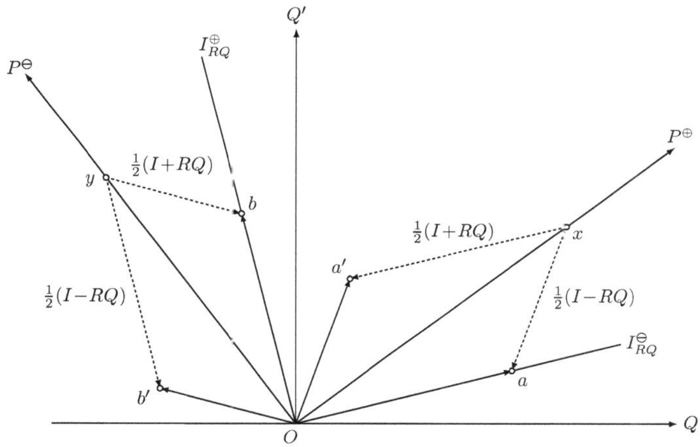

Fig. 9.8 Complementary images generated by linear mappings

are involved. The boundary data projections $Q^{\pm}$ are not involved; their place being taken by the two linear mappings, $\frac{1}{2}(I \pm RQ)$. The mappings are used to create two complementary images a, a' for every point x on the $P^{\oplus}$ axis, and two complementary images b, b' for every point y on the $P^{\ominus}$ axis, as shown in Fig. 9.8. The original points can be reconstructed from their images; $x = a + a'$ and $y = b + b'$.

The solution is initiated using two arbitrary starting points f_0^+ and f_1^- each on one of the two axes $P^{\oplus}$ and $P^{\ominus}$, notionally obtained in Fig. 9.9 by projection from a first guess of the joint interface function f, not shown off the top of the figure.

The two starting points are not on an equal footing. One is taken as an older estimate from which the other is derived. Each in turn is updated from the other. Here, f_0^+ is taken as the older estimate, and f_1^- is taken as the newer. The other order can be used if desired.

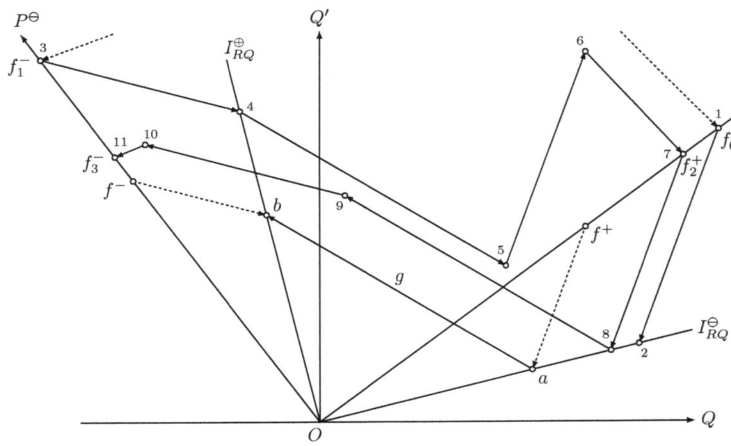

Fig. 9.9 General solution as geometric method

The projection to update the old f^+ from the new f^- is:

$$f_{2n}^{+} = P^{++}\left[\tfrac{1}{2}(I + RQ)f_{2n-1}^{-} - g + \tfrac{1}{2}(I + RQ)f_{2n-2}^{+}\right] \tag{9.16}$$

and the projection to update the (now) old f^- from the new f^+ is:

$$f_{2n+1}^{-} = P^{--}\left[\tfrac{1}{2}(I - RQ)f_{2n}^{+} + g + \tfrac{1}{2}(I - RQ)f_{2n-1}^{-}\right] \tag{9.17}$$

One cycle of operations, using Eqs. 9.16 and 9.17 with $n = 1$, is illustrated in Fig. 9.9 and listed in Table 9.1 with reference to the points labelled numerically in the figure. The step from 1 to 2 (coming in this case from the initial estimate but normally from the previous cycle) is used in the opposite sense to make the step from 5 to 6, and the step from 3 to 4 is used in the opposite sense to make the step from 9 to 10.

9.3.3 Computational Effort

The computational procedure to calculate the transmitted and reflected fields at all points on the interface is specified by Eqs. 9.16 and 9.17 and listed in Table 9.1. The starting point can be chosen as any vector which gives an estimate of the joint interface function f. A choice of zero is acceptable.

The computational effort is dominated by the Hardy projections, which require an integral over the entire interface. If the interface is subdivided into N elements, and the integral is evaluated as the sum of values calculated for each element, then

Table 9.1 Computational procedure for updating estimate of solution

Update	Steps	Figure 9.9
$f_2^+ :=$	$\frac{1}{2}(I + RQ)f_1^-$	3 → 4
	$-g$	4 → 5
	$+\frac{1}{2}(I + RQ)f_0^+$	5 → 6
	P^{++}	6 → 7
$f_3^- :=$	$\frac{1}{2}(I - RQ)f_2^+$	7 → 8
	$+g$	8 → 9
	$+\frac{1}{2}(I - RQ)f_1^-$	9 → 10
	P^{--}	10 → 11

The steps are shown graphically in Fig. 9.9 with labels as given in the rightmost column

the computational effort is proportional to the number of elements N. If the solution is required at one point on each element then the total computational effort is proportional to $N \times N = N^2$. This effort needs to be repeated for every iteration.

Once the solution $f^\pm$ has been obtained, calculating the field at one point away from the interface is performed using the Cauchy extension. Like the Hardy projection the Cauchy extension also requires an integral over the entire interface, with computational effort proportional to N. In order to calculate a complete map of the field at N points on a surface away from the interface requires N^2 operations from the Cauchy extension. This effort needs to be made only once.

Conventional techniques require some procedure equivalent to matrix inversion or the solution of a linear system of size N. That entails computational effort proportional to N^3. Accelerated techniques with higher speed but lower numerical stability can be constructed. In principle the computational effort can be reduced to slightly less than $N^{2.5}$ [50], although the algorithms are not practical [51] due to the very large (unstated) constant of proportionality. Currently the lower bound on the computational effort for any practical algorithm is about $N^{2.8}$ [52].

For some computational techniques, such as the Method of Moments [53–56], it is necessary before matrix inversion to use a time consuming numerical double integral to calculate the values of the individual matrix elements. For the iterative technique here, there is no such need. That results in a faster solution. Furthermore, for the iterative technique here the amount of memory storage required grows linearly with the size of the problem N, because there is never any need to store the elements of the matrix. In comparison, for the Method of Moments the matrix is stored, and the storage grows N^2 as the square of the size of the problem. For a given computer the Method of Moments is therefore always limited to solving smaller problems than the technique described here.

9.3.4 Joint Interface Function

The joint interface function f can be calculated from the values of the fields $f^{\pm}$, as depicted in Fig. 9.10. A first estimate f_1 of the joint interface function is refined using projections:

$$\begin{cases} P_1(x) = (I - P^{++})(x) + f^+ = P^{-+}(x) + f^+ \\ P_2(x) = (I - P^{--})(x) + f^- = P^{+-}(x) + f^- \end{cases} \tag{9.18}$$

to provide successive approximations:

$$\begin{cases} f_{2n} \quad = P_1(f_{2n-1}) \\ f_{2n+1} = P_2(f_{2n}) \end{cases} \tag{9.19}$$

as described in Sect. 3.5. Any choice of initial estimate, such as $f^{\pm}$ or zero, is acceptable.

The projection $P_1(x)$ maps a point x in the direction parallel to $OP^{\oplus}$ onto the line through f and f^+. The projection $P_2(x)$ maps a point x in the direction parallel to $OP^{\ominus}$ onto the line through f and f^-.

Repeated applications of P_1 and P_2 provide successive approximations f_n to the solution f. The amount of computational effort in finding the value of f is similar to that in first finding $f^{\pm}$.

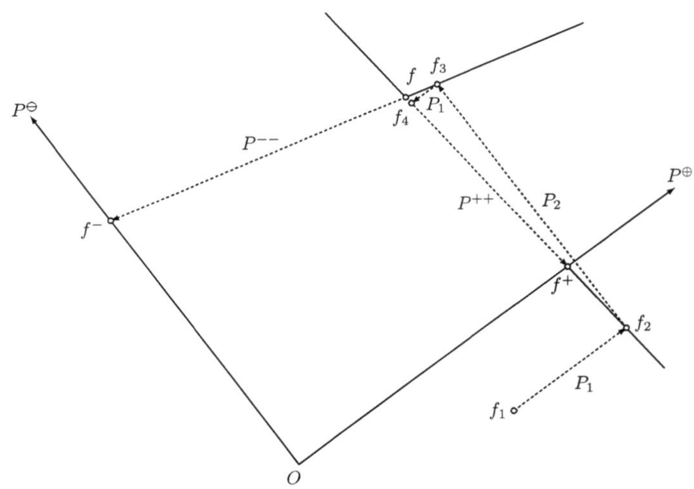

Fig. 9.10 Geometric solution for joint interface function f from initial estimate f_1

9.4 Convergence and Speed

In all of the methods of Sects. 9.1–9.3 the infinite summations or products are terminated at some finite value $0 < j < \infty$, and the result obtained at that stage is taken as the best estimate for the solution $f^{\pm}$. It is anticipated, as in Gauss-Seidel, that the solution is attained to sufficient accuracy before too many terms have been calculated. If the number of terms calculated exceeds the number of independent variables N taken to represent the electromagnetic field over the surface of the scatterer, then none of these techniques is faster than a full matrix inversion.

Chantaveerod [48] demonstrates convergence in about 30 iterations for metallic scatterers. The rate of convergence and accuracy for the method described in Sect. 9.3.2 is given in Sect. 10.3 for a wide variety of materials. Typically 30–40 iterations are sufficient. Taking 40 as representative, the iterative technique is expected to be faster than one based on matrix inversion when $40N^2 < N^{2.8}$, or $N > 100$. In most practical problems N exceeds 100 by a large margin. In most cases the method here is therefore expected to be significantly faster than techniques based on straightforward matrix inversion.

9.5 Exercises

Q1. Verify that $P_1 x = Q^+ x + g$ in Sect. 9.1 is a projection operator by showing that $P_1^2 = P_1$.

Q2. Verify that $P_1 x = Q^- x + h$ in Sect. 9.2 is a projection operator by showing that $P_1^2 = P_1$.

Q3. Show that the space-time scaling operator $R_{\alpha\beta}$ is neither a projection nor reflection operator.

Q4. Show that the space-time scaling operator $R_{\alpha\beta}$ is linear, as in Eq. 9.15.

Q5. Implement the method of solution for a perfectly transmitting cube as described in Sect. 9.2 by either geometric series, Horner's nested expansion, or alternating projections. Check the solution for the transmitted field against the incident field.

Q6. Implement the method of solution for a perfectly conducting cube as described in Sect. 9.1 by either geometric series, Horner's nested expansion, or alternating projections. Check the solution for the reflected field against the measured data from [41–43].

Q7. Implement the method of solution for a cube of arbitrary material properties as described in Sect. 9.3.2. Verify the solution against Exercises 5 and 6 in the cases $\Gamma = 0$ and $\Gamma = -1$ for perfect transmission and reflection.

Q8. Show that the operator P_1 in Eq. 9.19 is indeed a projection operator, and that it maps points onto the line through f and f^+.

Q9. Show that the operator P_2 in Eq. 9.19 is indeed a projection operator, and that it maps points onto the line through f and f^-.

Q10. Redraw Fig. 9.10 using f^- as the initial estimate f_1.

Part III
Demonstration

Chapter 10
Examples

The method described in Sect. 9.3 solves the general scattering problem in three dimensions for lossless and lossy dielectrics and conductors. Lossless dielectrics and perfect conductors are special cases of the general solution.

The method is geometric in a Banach space of two dimensions, and can be cast into affine geometry in the plane as in Fig. 9.9. One dimension of the solution relates to the need to adhere to Maxwell's equations of electromagnetism in regions of uniform material properties without sources, cast in the form of Hardy projections. The other dimension relates to the need to match the fields according to physical constraints at the interfaces where the boundaries of different regions meet, cast in the form of boundary data projection operators.

The method applies equally well from one to three spatial dimensions with appropriate selection of the Dirac operator and the Cauchy Kernel.

In the case of one spatial dimension, solutions relate to physical problems involving either transmission lines with discontinuities, or plane waves penetrating planar barriers. These problems are taken here as useful test cases for the method. Solutions are already known by independent means, which allows results to be checked rigorously for accuracy. In addition part of the algebra, in particular the Cauchy integral, can be carried analytically so that the Hardy projections reduce to formulae which can be evaluated directly and without approximation. That makes the implementation efficient and allows a thorough experimental study of the method under a wide variety of different conditions.

A succinct description of the one-dimensional test problem is given in Sect. 10.1. The problem is fully formulated in Sect. 10.2, drawing on the theory in part II. Results from implementation of the solution are given in Sect. 10.3.

A. Seagar, *Application of Geometric Algebra to Electromagnetic Scattering*,
DOI 10.1007/978-981-10-0089-8_10

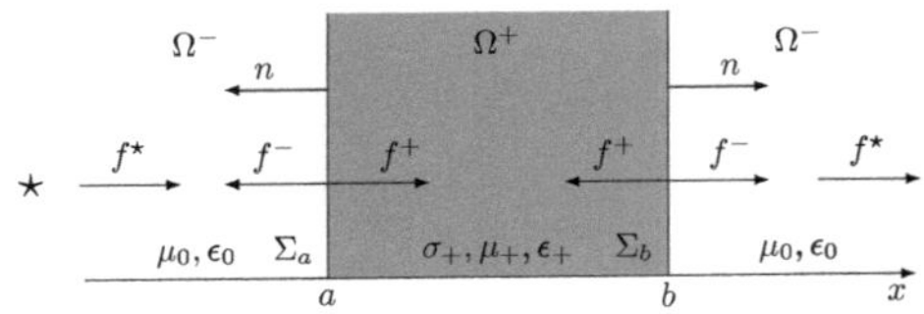

Fig. 10.1 Radiation from distant source ★ is transmitted and reflected at surface Σ of scatterer Ω^+ embedded in free space Ω^-

10.1 Test Case

Figure 10.1 shows a planar object Ω^+ of finite thickness $b-a \geq 0$ in one dimension, x, and infinite size in the two orthogonal dimensions, with uniform electrical conductivity σ_+, permittivity ϵ_+ and permeability μ_+. The object is embedded in a region of free space Ω^- with conductivity $\sigma_- = 0$, permittivity $\epsilon_- = \epsilon_0$ and permeability $\mu_- = \mu_0$.

A linearly polarised plane wave of electromagnetic radiation $f^\star$ from a distant source ★ strikes the surface $\Sigma = \Sigma_a \cup \Sigma_b$ of the object Ω^+ along the direction of the normal to the surface, n. Part of the wave f^+ penetrates and is admitted into the object, and part of the wave f^- fails to penetrate and is scattered away from the object. The total field measured in free space Ω^- is constructed from the sum of the incident plane wave $f^\star$ and the scattered (or reflected) wave f^-. Inside the object the total field is constructed from the admitted (or transmitted) wave f^+ alone.

Tests involve varying the values of the quantities σ_+, μ_+ and ϵ_+ to cover all types of materials (as listed in Table B.1 in Appendix B) for a chosen value of thickness $b-a$ and angular frequency ω.

10.2 Formulation

Formulation of the one-dimensional problem involves deriving the Hardy projections, for use in the computational procedure listed in Table 9.1. The process starts by recasting the Dirac operator (Sect. 5.4) into one dimension, and establishing a current source which produces a plane wave of electromagnetic radiation. Maxwell's equations are written in Clifford form for this particular current source, and also for a planar scalar point source. The one-dimensional version of the Bessel potential (Eq. 7.7) is used to construct the one-dimensional fundamental solution (Sect. 7.2) for the point source, from which the general field solution for a homogeneous region is obtained. The incident field $f^\star$ is based on that solution.

The one-dimensional Cauchy kernel (Sect. 7.3.1) is constructed from the fundamental solution. From that, closed form algebraic formulae are determined for the Cauchy extension (Sect. 7.3.2) for all regions in space away from the surface, and closed form algebraic formulae are developed for the Cauchy integral (Sect. 7.4.1) at all points on the surface. That gives directly the Hardy projections (Chap. 8), which are then used in the computational procedure listed in Table 9.1.

10.2.1 Dirac Operator

With material properties and fields constant in two directions, derivatives in those directions give zero value. The Dirac and k-Dirac operators (Sect. 5.4) in this situation can be written using only the components which produce non-zero values:

$$\mathcal{D} = \frac{\partial}{\partial x} e_1 - \frac{i}{c} \frac{\partial}{\partial t} e_0 \tag{10.1}$$

and:

$$\mathcal{D}_k = \frac{\partial}{\partial x} e_1 + k e_0 \tag{10.2}$$

10.2.2 Source

A uniform sinusoidal current source $\mathcal{S}_k$ in the form of an infinitesimally thin sheet J_y Ampères per metre is established within the plane $x=0$ flowing in the $-y$ direction:

$$\mathcal{S}_k = \delta(x - 0)[-J_y e_2] \tag{10.3}$$

as shown in Fig. 10.2. This particular source is chosen in order to generate a planar electromagnetic field propagating away from the origin along the x axis in both directions (see Sect. 10.2.6).

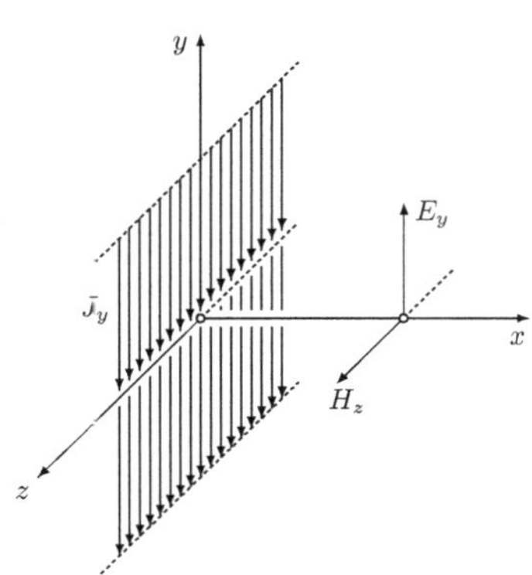

Fig. 10.2 Current sheet as source for electromagnetic field

10.2.3 Maxwell's Equations

The electromagnetic field $\mathcal{F}_k$ generated by the source satisfies Maxwell's equations (see Sect. 5.3):

$$\mathcal{D}_k \mathcal{F}_k = \mathcal{S}_k = \delta(x-0)[-J_y e_2] \tag{10.4}$$

10.2.4 Fundamental Solution

The solution for the electromagnetic field $\mathcal{F}_k$ is constructed here, as in Sect. 7.2, from the fundamental solution F_k for the scalar point source $\delta(x-0)$, under the influence of the one-dimensional k-Dirac operator:

$$\mathcal{D}_k F_k = \delta(x-0) \tag{10.5}$$

Multiplying by $-J_y e_2$ and comparing to Eq. 10.4 gives the electromagnetic field in terms of the fundamental solution:

$$\mathcal{F}_k = F_k[-J_y e_2] \tag{10.6}$$

10.2.5 Bessel Potential

The fundamental solution F_k to the k-Dirac operator is itself constructed from either of the two fundamental solutions $B_{\pm k}$ of the Helmholtz operator $H_k = -\mathcal{D}_k^2$:

$$[-\mathcal{D}_k^2] B_{\pm k} = \delta(x-0) \tag{10.7}$$

The Bessel potential $B_{\pm k}$ (see Sect. 7.2) is given in one dimension by [57, 58]:

$$B_{\pm k}(\mathbf{r}) = \pm \tfrac{1}{2ik} e^{\pm ikr} \tag{10.8}$$

where $\mathbf{r} = x e_1$ and $r = |\mathbf{r}| = |x|$. Here, as in Sect. 7.2.1, to obtain a field flowing away from the source it is the solution with the negative wavenumber, $-k$, which is used.

Comparing Eqs. 10.5 and 10.7 and using Eq. 10.8 gives the fundamental solution in one dimension:

$$F_k = -\mathcal{D}_k B_{-k} = -(\tfrac{\partial}{\partial x} e_1 + k e_0)\left[-\tfrac{1}{2ik} e^{-ik|x|}\right] = -\tfrac{1}{2}(\hat{\mathbf{r}} + i e_0) e^{-ikr} \tag{10.9}$$

where $\hat{\mathbf{r}}$ is a unit vector in the direction of $\mathbf{r}$.

10.2.6 Electromagnetic Field

Using the fundamental solution 10.9, the solution for the field is obtained from Eq. 10.6 in terms of the current source:

$$\mathcal{F}_k = -\tfrac{1}{2}(\hat{\mathbf{r}} + ie_0)e^{-ikr}[-J_y e_2] = \tfrac{1}{2}J_y(\hat{\mathbf{r}}e_2 + ie_0e_2)e^{-ikr}$$
$$= \begin{cases} \tfrac{1}{2}J_y(+e_3\sigma - ie_2e_0)e^{-ikx} & x>0 \\ \tfrac{1}{2}J_y(-e_3\sigma - ie_2e_0)e^{+ikx} & x<0 \end{cases} \quad (10.10)$$

Comparison to the general form for the electromagnetic field (see Sect. 5.3):

$$\mathcal{F}_k = \sqrt{\mu}(H_xe_1+H_ye_2-H_ze_3)\sigma - i\sqrt{\epsilon}(E_xe_1+E_ye_2+E_ze_3)e_0 \quad (10.11)$$

shows that here for the sheet current source in the one-dimensional problem $H_x = H_y = E_x = E_z = 0$ and $H_z = \pm\frac{1}{2\sqrt{\mu}}J_ye^{\pm ikx}$ and $E_y = \frac{1}{2\sqrt{\epsilon}}J_ye^{\pm ikx}$.

The first line of Eq. 10.10 is the Clifford-valued phasor of a plane wave $\frac{1}{2}J_y(e_3\sigma - ie_2e_0)e^{i(\omega t-kx)}$ travelling in the positive x direction away from the origin. By applying the right hand rule it is seen that the direction of the magnetic field immediately on the right hand side of the origin, $+\frac{1}{2\sqrt{\mu}}J_ye_3$, is consistent with the direction chosen for the current source. Applying Poynting's theorem $\mathbf{P} = \mathbf{E} \times \mathbf{H}$ confirms that the electric and magnetic fields are consistent with propagation away from the origin. Inspecting the exponent $\omega t - kx$ (as in Sect. 7.2.1) shows that the velocity dx/dt is positive.

The second line of Eq. 10.10 is the Clifford-valued phasor of a plane wave $\frac{1}{2}J_y(-e_3\sigma - ie_2e_0)e^{i(\omega t+kx)}$ travelling in the negative x direction away from the origin. By applying the right hand rule it is seen that the direction of the magnetic field immediately on the left hand side of the origin, $-\frac{1}{2\sqrt{\mu}}J_ye_3$, is consistent with the direction chosen for the current source. Applying Poynting's theorem $\mathbf{P} = \mathbf{E} \times \mathbf{H}$ confirms that the electric and magnetic fields are consistent with propagation away from the origin. Inspecting the exponent $\omega t + kx$ shows that the velocity dx/dt is negative.

For the test case here a sheet current source $J_y = 2$ Ampères per metre placed some distance far to the left of the origin is taken as a source $\star$ to establish the incident electromagnetic plane wave $f^\star e^{i\omega t}$ travelling in the positive x direction, as in Fig. 10.1:

$$f^\star e^{i\omega t} = (e_3\sigma - ie_2e_0)e^{i(\omega t-kx)} \quad (10.12)$$

10.2.7 Cauchy Kernel

The Cauchy extension and Cauchy integral in one dimension follow the general formulae of Sects. 7.3.2 and 7.4.1, along with the one-dimensional Cauchy kernel E_k, generated as in Sect. 7.3.1 from the fundamental solution F_k:

$$E_k(\mathbf{r}) = -F_k(-\mathbf{r}) = -\tfrac{1}{2}(\hat{\mathbf{r}} - ie_0)e^{-ikr} \tag{10.13}$$

10.2.8 Cauchy Extension

The Cauchy extension, from Sect. 7.3.2:

$$Cf(\mathbf{r} \notin \Sigma) = \int_{\Sigma} E_k(\mathbf{s} - \mathbf{r})n(\mathbf{s})f(\mathbf{s})d\sigma(\mathbf{s}) \tag{10.14}$$

is separated into two components A, B, one for each part Σ_a, Σ_b of the boundary Σ:

$$Cf(x \notin \Sigma) = A + B = \int_{\Sigma_a} E_k((a-x)e_1)n(a)f(a)d\sigma(a) + \int_{\Sigma_b} E_k((b-x)e_1)n(b)f(b)d\sigma(b) \tag{10.15}$$

The result of evaluating each of the two integrals A and B for points on either side of the two interfaces at $x = a$ and $x = b$ is shown in Table 10.1. Terms in the integrals of the form $\int_{\Sigma_a} 1d\sigma(a)$ represent the area of the interfaces. For convenience, the whole problem is rescaled to reduce their values to unity.

Applying the integrals in Table 10.1 to construct the Cauchy extension in the three regions to the left, in the middle, and to the right of the object Ω^+ in Fig. 10.1, gives:

$$-f^-(x) = -\tfrac{1}{2}(1 + ie_0e_1)e^{-ik(a-x)}f(a) + \tfrac{1}{2}(1 + ie_0e_1)e^{-ik(b-x)}f(b) \quad ;\textit{left} \quad x<a<b \tag{10.16}$$

$$+f^+(x) = +\tfrac{1}{2}(1 - ie_0e_1)e^{-ik(x-a)}f(a) + \tfrac{1}{2}(1 + ie_0e_1)e^{-ik(b-x)}f(b) \quad ;\textit{middle} \quad a<x<b \tag{10.17}$$

$$-f^-(x) = +\tfrac{1}{2}(1 - ie_0e_1)e^{-ik(x-a)}f(a) - \tfrac{1}{2}(1 - ie_0e_1)e^{-ik(x-b)}f(b) \quad ;\textit{right} \quad a<b<x \tag{10.18}$$

Table 10.1 Integrals for the one-dimensional Cauchy extension

$x < a$ $\quad n = -e_1 \quad \hat{\mathbf{r}} = +e_1$	$x > a$ $\quad n = -e_1 \quad \hat{\mathbf{r}} = -e_1$
$A = \int_{\Sigma_a} -\frac{1}{2}(e_1 - ie_0)e^{-ik(a-x)}(-e_1)f(a)d\sigma(a)$ $= -\frac{1}{2}(1+ie_0e_1)e^{-ik(a-x)}f(a)$	$A = \int_{\Sigma_a} -\frac{1}{2}(-e_1 - ie_0)e^{-ik(x-a)}(-e_1)f(a)d\sigma(a)$ $= +\frac{1}{2}(1-ie_0e_1)e^{-ik(x-a)}f(a)$
$x < b$ $\quad n = +e_1 \quad \hat{\mathbf{r}} = +e_1$	$x > b$ $\quad n = +e_1 \quad \hat{\mathbf{r}} = -e_1$
$B = \int_{\Sigma_b} -\frac{1}{2}(e_1 - ie_0)e^{-ik(b-x)}e_1 f(b)d\sigma(b)$ $= +\frac{1}{2}(1+ie_0e_1)e^{-ik(b-x)}f(b)$	$B = \int_{\Sigma_b} -\frac{1}{2}(-e_1 - ie_0)e^{-ik(x-b)}e_1 f(b)d\sigma(b)$ $= -\frac{1}{2}(1-ie_0e_1)e^{-ik(x-b)}f(b)$

Note that these expressions are purely algebraic and require no form of integration, numerical or otherwise.

10.2.9 Cauchy Integral

The Cauchy integral (see Sect. 7.4.1):

$$C_\Sigma f(\mathbf{r} \in \Sigma) = 2\,\text{p.v.} \int_\Sigma E_k(\mathbf{s}-\mathbf{r}) n(\mathbf{s}) f(\mathbf{s}) d\sigma(\mathbf{s}) \tag{10.19}$$

is constructed where necessary from the limiting values of the Cauchy extension for two points approaching each interface Σ_a, Σ_b from opposite sides:

$$\begin{aligned} C_\Sigma f(a) = 2 \Bigg\{ & \int_{\Sigma_b} E_k((b-a)e_1) n(b) f(b) d\sigma(b) + \tfrac{1}{2} \int_{\Sigma_a} E_k\left(\tfrac{1}{\infty - 1} e_1\right) n(a) f(a) d\sigma(a) \\ & + \tfrac{1}{2} \int_{\Sigma_a} E_k\left(\tfrac{1}{1-\infty} e_1\right) n(a) f(a) d\sigma(a) \Bigg\} \\ = {} & (1 + i e_0 e_1) e^{-ik(b-a)} f(b) - i e_0 e_1 f(a) \end{aligned} \tag{10.20}$$

and:

$$\begin{aligned} C_\Sigma f(b) = 2 \Bigg\{ & \int_{\Sigma_a} E_k((a-b)e_1) n(a) f(a) d\sigma(a) + \tfrac{1}{2} \int_{\Sigma_b} E_k\left(\tfrac{1}{\infty - 1} e_1\right) n(b) f(b) d\sigma(b) \\ & + \tfrac{1}{2} \int_{\Sigma_b} E_k\left(\tfrac{1}{1-\infty} e_1\right) n(b) f(b) d\sigma(b) \Bigg\} \\ = {} & (1 - i e_0 e_1) e^{-ik(b-a)} f(a) + i e_0 e_1 f(b) \end{aligned} \tag{10.21}$$

In the expressions the terms $\frac{1}{\infty-1}$ and $\frac{1}{1-\infty}$ represent respectively positive and negative values infinitesimally away from zero. The expressions which result are fully resolved, carry no singularity, and properly account for the discontinuity in the values of the field at the interface.

10.2.10 Hardy Projections

The Hardy projections (see Chap. 8) are calculated from the Cauchy integral:

$$P^{\pm} f = \tfrac{1}{2}(I \pm C_\Sigma) f \tag{10.22}$$

This gives:

$$\begin{cases} P^{+}f(a) = \frac{1}{2}(f(a) + C_{\Sigma}f(a)) = \frac{1}{2}((1 - ie_0e_1)f(a) + (1 + ie_0e_1)e^{-ik(b-a)}f(b)) \\ P^{-}f(a) = \frac{1}{2}(f(a) - C_{\Sigma}f(a)) = \frac{1}{2}((1 + ie_0e_1)f(a) - (1 + ie_0e_1)e^{-ik(b-a)}f(b)) \end{cases} \tag{10.23}$$

and:

$$\begin{cases} P^{+}f(b) = \frac{1}{2}(f(b) + C_{\Sigma}f(b)) = \frac{1}{2}((1 + ie_0e_1)f(b) + (1 - ie_0e_1)e^{-ik(b-a)}f(a)) \\ P^{-}f(b) = \frac{1}{2}(f(b) - C_{\Sigma}f(b)) = \frac{1}{2}((1 - ie_0e_1)f(b) - (1 - ie_0e_1)e^{-ik(b-a)}f(a)) \end{cases} \tag{10.24}$$

The projections P^{++} and P^{--} associated with the regions Ω^{+} and Ω^{-} respectively are obtained using the corresponding values of wavenumber $k = k^{+}$ and $k = k^{-}$, within the Cauchy kernel $E_k = E_{k^{\pm}}$ inside the Cauchy integral $C_{\Sigma} = C_{\Sigma}^{\pm}$, as in Sect. 8.3.

Taking Eqs. 10.23 and 10.24 with appropriate values of $k^{\pm}$ now provides all that is required to implement the solution to the scattering problem via the computational procedure listed in Table 9.1.

10.3 Results

The computational procedure has been applied for the 36 materials listed in Table B.1 at an angular frequency of $\omega = 4\pi \times 10^9$ Hz, for a scatter of thickness $b-a$ equal to ten wavelengths in free space. The materials are chosen to illustrate the full range of values the permittivity and conductivity might well be expected to cover. For tests under other conditions see [59].

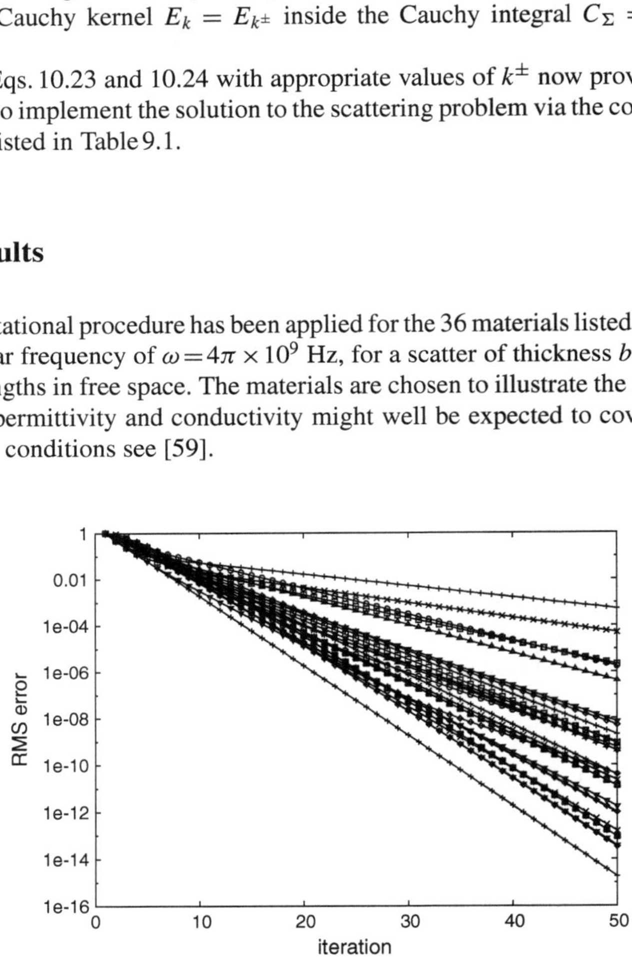

Fig. 10.3 Convergence of iterative method for different materials

Table 10.2 Rates of convergence and errors in solution after 50 iterations

Material	Error Δg_{50}	Rate $\Delta g_{50}/\Delta g_{49}$
Ethanol	5.398×10^{-4}	0.891
Water	5.056×10^{-5}	0.861
Gallium arsenide	2.153×10^{-6}	0.794
Soda-lime glass	1.769×10^{-6}	0.771
Silicon carbide	4.244×10^{-7}	0.756
Sapphire	7.080×10^{-9}	0.696
Acetone	5.124×10^{-9}	0.695
Lead oxide glass	1.989×10^{-9}	0.671
Germanium	9.115×10^{-10}	0.680
Acetic acid	8.082×10^{-10}	0.657
Sea water	5.596×10^{-10}	0.690
Silicon	4.023×10^{-10}	0.659
Diamond	3.891×10^{-10}	0.649
PZT	3.999×10^{-11}	0.694
Polystyrene	3.686×10^{-11}	0.608
Pyrex	2.164×10^{-11}	0.611
Polyethylene	1.270×10^{-11}	0.603
Quartz (synthetic)	1.265×10^{-11}	0.602
Quartz (natural)	1.265×10^{-11}	0.602
Octane	1.688×10^{-12}	0.588
Polypropylene	9.716×10^{-13}	0.570
Silica glass	9.134×10^{-13}	0.569
PET	1.548×10^{-13}	0.542
Bismuth	8.749×10^{-14}	0.511
Mercury	8.736×10^{-14}	0.511
Graphite	8.713×10^{-14}	0.510
Antimony	8.711×10^{-14}	0.510
Arsenic	8.711×10^{-14}	0.510
Gold	8.710×10^{-14}	0.510
Copper	8.705×10^{-14}	0.510
Silver	8.705×10^{-14}	0.510
Aluminium	8.702×10^{-14}	0.510
Platinum	8.689×10^{-14}	0.510
Polycarbonate	3.679×10^{-14}	0.515
PVC	3.310×10^{-14}	0.511
Air	1.842×10^{-15}	0.497

The procedure was implemented in the C programming language. The programming code is listed in Appendix C.

Following the methodology adopted in [59], the results are given in Fig. 10.3 in terms of the RMS value of the difference Δg between the actual data g and that given by the current estimates of the fields:

$$\Delta g_n = \|\tfrac{1}{2}(I+RQ)f^-_{2n+1} - \tfrac{1}{2}(I-RQ)f^+_{2n} - g\| \tag{10.25}$$

The procedure is started with $n=1$ using first estimates of the fields $f^+_0 = f^-_1 = 0$, and continued until fifty iterations have been completed.

As in [59], the algorithm converges here for all types of materials. The convergence is linear, in the manner of a geometric series. For the materials tested, the rate of convergence $\Delta g_{50}/\Delta g_{49}$ varies from the slowest value of 0.891 to the fastest value of 0.497, as listed in Table 10.2. The horizontal lines separating groups of materials in the table correspond to the seven distinct empty bands separating clusters of traces in Fig. 10.3.

Faster convergence is typically observed for lower values of the contrast in permittivity ϵ_+/ϵ_-, and for higher values of losses. Difficulties with other algorithms have been observed when dielectric materials have low values of contrast [60, 61]. Here the algorithm performs in that case, for air with relative permittivity 1.0006, the fastest of all.

When the contrast in permittivity is high and losses are low, as for ethanol and water, slower convergence is typically observed. If there is sufficiently high loss, as for lead zirconate titanate (PZT) with $\tan\delta = 0.015$, even a high value of contrast ($\epsilon/\epsilon_0 = 1800$) converges rapidly. When the same loss is attributed to ethanol and water, they (i.e. the slightly modified materials) converge in 50 iterations with errors of 8.080×10^{-10} and 6.110×10^{-10} respectively, and with rates of convergence 0.678 and 0.689. Introducing some loss effectively damps any resonance which may otherwise slow down the solution.

10.4 Exercises

Q1. Replacing the one-dimensional Green's function with the two-dimensional Green's function, modify the code in Appendix C to produce solutions for problems with material properties constant in one dimension and waves propagating perpendicular to that direction.

Q2. Replacing the one-dimensional Green's function with the three-dimensional Green's function, modify the code in Appendix C to produce solutions for problems with material properties which vary in any direction.

Part IV
Contemplation

Chapter 11
Perspectives

11.1 Genesis

Conventional techniques for solving problems involving electromagnetic scattering and the technique described here, are both based on mathematics developed in the nineteenth century as depicted in Fig. 11.1. The conventional techniques, in the box on the lower right, are linked by arrows to the individuals on whose work they rely. Individuals whose work underlies the technique described in part II of this book, are linked to the box on the lower left of the figure.

The conventional approach is to largely ignore the work of Cauchy, Grassmann and Clifford. The alternative approach is to discard work of Green, Gibbs and Heaviside. Discarding the work of the latter group may seem somewhat sacrilegious, but following it too religiously could also be an error of judgement.

During the nineteenth century, when all of these different mathematical tools were being developed, there was discussion, at times heated, over which were the best tools to use in the formal description and study of electromagnetic and other physical phenomena. By the twentieth century the contest had largely been won by vector calculus, as promoted by Gibbs and Heaviside. The adoption of that approach must have at least in part been influenced in a Darwinian fashion, with all of the proponents of the alternative having, by that time, departed. However, the selection was based on the better fitness of the individuals, and not necessarily the technique.

The name "*Clifford-Cauchy-Dirac* (CCD) technique" is adopted here as in [59], for the more recently developed of the two approaches, with the intention of emphasising the fundamental distinction between them. The work of Cauchy and Clifford are essential to the new method. It is the combination of the two which supports generalisation of the Cauchy integral to multiple dimensions [10]. Dirac is included for the use of the Dirac operator [31], which is effectively a modern form of the quaternion operator first used by Maxwell, discarded by Gibbs and Heaviside, and reintroduced by Dirac in the twentieth century in a different context [32, 33].

A. Seagar, *Application of Geometric Algebra to Electromagnetic Scattering*,
DOI 10.1007/978-981-10-0089-8_11

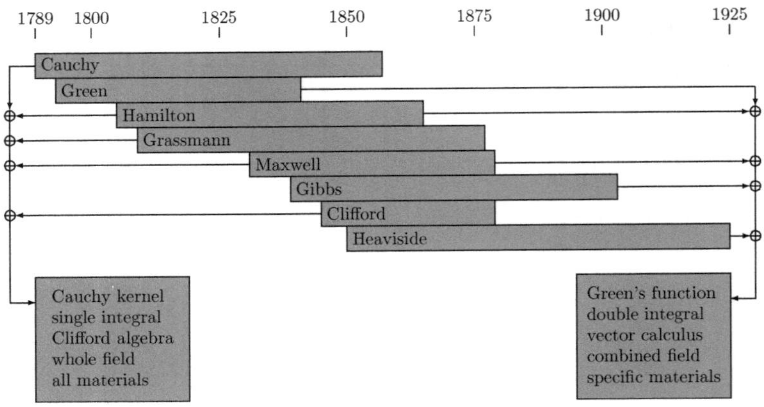

Fig. 11.1 Genesis of the Clifford-Cauchy-Dirac (CCD) and conventional techniques

Techniques for solving scattering problems formulated in terms of Green's functions, double integrals, vector calculus and the combined field (CFIE) approach lead through implementation by the method of moments (MoM) to what promises to be a general solution. Application of this approach to perfectly conducting objects works well [62]. However, application to lossy conductors and dielectrics [63, 64] is for some reason more difficult. It appears that no single algorithm comes anywhere close to covering all cases with adequate accuracy [60, 61, 65, 66].

The CCD method, formulated instead in terms of the Cauchy kernel, single integrals, Clifford algebra and a whole field approach, also promises a general solution. For this alternative method, based on the results in Chap. 10 and [59], the promise in one dimension at least is properly honoured.

Currently, in three dimensions the new approach has only been verified for perfect conductors [48]. The behaviour of the solution for that special case in both one and three dimensions is effectively the same. An inference may be drawn that the general method here also behaves the same in one and three dimensions for all other materials, although that is yet to be verified.

11.2 Two and Three Dimensions

An implementation can be constructed for two or three dimensions using as a framework the code in Appendix C. The overall structure of the algorithm remains the same, being unaffected by the number of dimensions.

Three modifications are necessary:

- increase the number of elements (tiles) on the surface from 2 to N,
- replace the code for the one-dimensional Cauchy kernel by code for the two or three-dimensional Cauchy kernel, and

- recode routines which operate on fields over the zero-dimensional boundary of a one-dimensional region to operate instead on fields over the one-dimensional boundary of a two-dimensional region or over the two-dimensional boundary of a three-dimensional region.

The first two modifications are trivial.

For the third modification, recoding the linear operators $\frac{1}{2}(I \pm RQ)$ is straightforward because they operate locally, at isolated points. More effort is required to recode the Hardy projections $P^{\pm\pm}$, which operate globally, requiring a contribution from the field at every point on the surface to calculate the result at just one point. Care must be taken to properly accommodate the integrable singularity at the point where the result is to be determined.

11.3 Multiple Scatterers

The method for solving scattering problems in part II is formulated for the case where there is only one scatterer. If there are multiple scatterers it is necessary to extend and reformulate the problem accordingly. Figure 11.2 shows two cases of particular interest.

In Fig. 11.2a a wave scatters from the surfaces of two objects in parallel, and in Fig. 11.2b the wave scatters from two objects in series. These two cases are entirely analogous to series and parallel combinations of transmission lines. Parallel and series combinations are the two fundamental means of construction by which any arbitrary network of components (i.e. circuit) can be constructed. The shape of the scatterers for any specific problem is irrelevant since the method used here automatically incorporates that into the behaviour of the boundary data projection operators and the Hardy projection operators.

The scattering problem for regions of different material properties, as in Fig. 11.2, has been demonstrated at zero frequency (wavenumber $k = 0$) in the case where the electrical conductivity σ dominates the intrinsic impedance η [67]. In that work the equations are treated as second order partial differential equations and conventional methods used, not the methods described in part II. However the issue of dealing with multiple scatterers of the type shown in Fig. 11.2 has been successfully addressed.

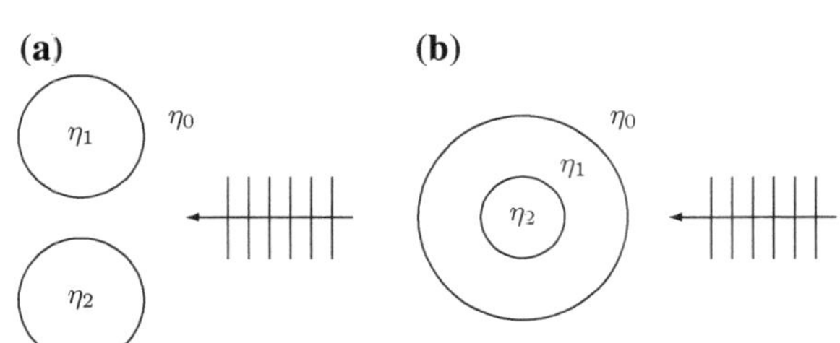

Fig. 11.2 Scattering from two objects (**a**) in parallel, (**b**) in series

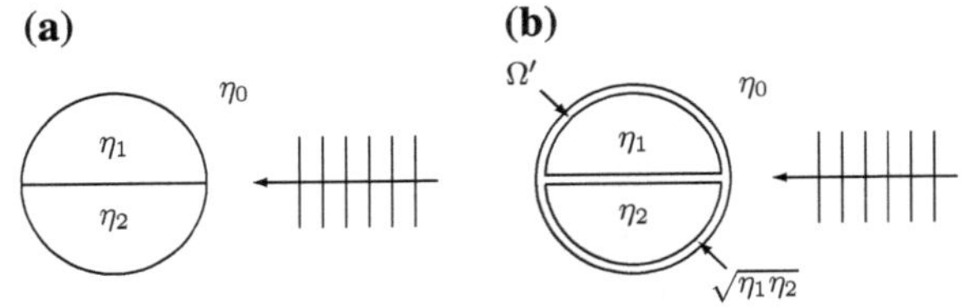

Fig. 11.3 Three regions (**a**) with common three-way junction, (**b**) treated as four regions with three-way junction eliminated

A third case of interest is shown in Fig. 11.3a where three regions of different material properties meet at a common point.

In principle this is simply the series and parallel combination of Fig. 11.3b, where the region Ω' has shrunk in size arbitrarily close to zero. Provided that methods to handle the parallel and series cases of Fig. 11.2 can be constructed, it then follows that all cases can be solved. See for example the figures in [67].

When multiple scatterers are involved the size of the linear system grows in proportion to the total number of distinct interfaces where adjacent regions meet. For example, in Fig. 11.3a there are three interfaces; between η_0 and η_1, η_0 and η_2, and η_1 and η_2. Although there is an extra region Ω' in Fig. 11.3b, there are still only three interfaces. A different pair of Hardy projections is required for each interface.

11.4 Topology and Graphs

The complexity of a problem involving multiple scatterers can be determined by examining the topological structure of the problem rather than its geometrical structure. The topological structure is obtained by replacing every region by a single point and every interface by a link between a pair of these points. That produces a framework of interconnected points, know in mathematical terminology as a *graph* (see Fig. 11.4).

The graph retains, as links, the connective relationships between regions of different material properties. It does not retain the shapes of the original surfaces, regions

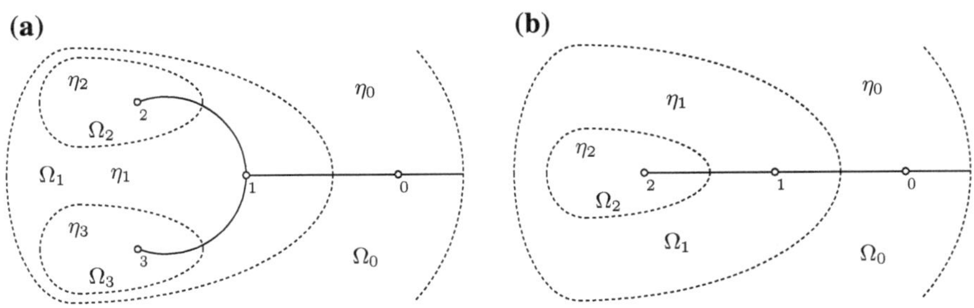

Fig. 11.4 Transformation of scattering problems into equivalent graphs with links in (**a**) parallel, (**b**) series. Interfaces (*dotted*) are replaced by links (*solid lines*), and regions Ω are replaced by points (○)

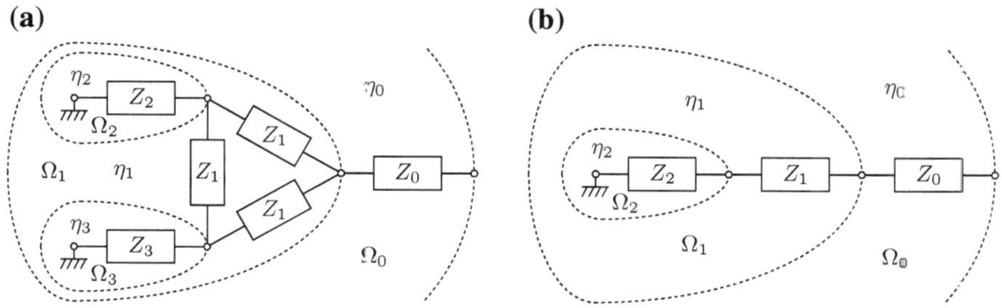

Fig. 11.5 Collapse of scattering problems onto equivalent circuits with components in (**a**) parallel, (**b**) series. Interfaces (*dotted*) reduce to nodes (∘), and regions Ω reduce to components Z

or interfaces. For each link two equations are required, one for a wave travelling in one direction and one for a wave travelling in the opposite direction. Each equation is formulated in terms of Hardy projections. The number of equations required to solve the full problem with multiple scatterers is governed by the number of links in the equivalent graph. It is indifferent to the shapes of objects.

11.5 Equivalent Circuits

There is a direct one-to-one equivalence between the graph of a scattering problem and an electrical circuit. Each point in the graph plays the role of component in the circuit, and each link in the graph represents a wire connecting two components. The effort required for the solution of the problem depends on the number of links in the graph or the number of wires in the circuit.

In the case of a circuit, the equations at the nodes involve field variables which occupy only one dimension. For the scattering problem in three dimensions, the equations at the nodes (interfaces) involve variables with the number of dimensions equal to the number of elements employed to represent the interface. In terms of a matrix representation, every equation leads to a block of rows and columns, with non-zero entries only where rows and columns are coupled by a common node. In circuits, these matrices are often quite sparse.

Some components have more than two terminals, as for Z_1 in Fig. 11.5a, and can be represented in more than one equivalent form. See for example exercise 5.

11.6 Electromagnetism

In the case of electromagnetism, it seems clear that not only does the Clifford-Cauchy-Dirac technique provide a viable alternative method to conventional techniques, it provides one which in many ways is superior. By comparing the fundamental similarities and differences between the CCD technique and the Method of Moments [53–56] it seems likely that the Clifford-Cauchy-Dirac technique offers very real and significant advantages in uniformity, complexity, speed, storage, stability, consistency and accuracy. These advantages have yet to be demonstrated fully in three dimensions.

11.7 Beyond Electromagnetism

The methods described in part II are focussed on solving particular kinds of physical problems involving the scattering of electromagnetic waves, using Maxwell's equations as a classical mathematical representation of that phenomenon. Maxwell's equations are in the form of certain first order differential equations, which can be further interpreted as square roots of the (second order) Laplace, Helmholtz and wave equations. Given that these and similar equations are found to represent many and varied physical phenomena, it seems likely that the method adopted here can be recast and applied also to those other phenomena. This seems to be particularly the case where elliptic operators are involved, at least in those situations where their 'square roots' can be shown to be properly defined and well behaved (see [68–70]). It should be possible develop methods of solution for these other physical phenomena in a similar style as is done here for electromagnetism.

11.8 Exercises

Q1. Develop a method of solution for scattering from two objects in parallel as in Fig. 11.2a.

Q2. Develop a method of solution for scattering from two objects in series as in Fig. 11.2b.

Q3. Modify the result from question 1 to accommodate the case depicted in Fig. 11.3.

Q4. Using the results from question 2 and 3 develop a general solution for any number of scatterers in any equivalent circuit.

Q5. Find the values of the impedances Z_1' for which the circuit below is an electrically identical replacement for the three impedances Z_1 in Fig. 11.5a.

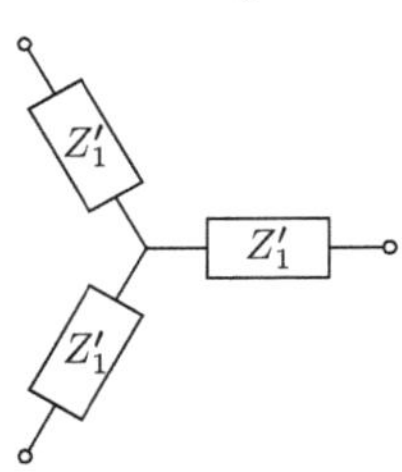

Q6. For a physical phenomenon of your choice as described by an elliptic operator, develop a method of solution using the square root of that operator for a problem involving multiple regions of different but uniform material properties meeting at interfaces.

Appendix A
Reduce Code for Machine Computation

The source code listed here is taken verbatim from a working suite of computer routines. It operates on Clifford numbers with either numerical or symbolic coefficients. Translation to the LISP language is relatively straightforward, since REDUCE is built on either standard LISP or some later variant. However, the facility for symbolic calculation is then lost. Translation to the C language takes more effort but can offer advantages in terms of speed.

For the electromagnetic application, routines as general as these are wasteful. In that case it is normally known that the terms in a product are either vectors or bivectors of known dimension, and it is much more computationally efficient to write a suite of routines to cover those particular cases, never attempting to multiply components which are known a priori to be zero.

The definitions of the products here are taken from Tables 2.4 and 2.5. The Clifford conjugations shown in those tables on one or other of the multiplicands for the inner-style products are invoked as the first operation within the computer routines. For two arguments a, b these routines therefore behave as $a \vee b$, $a \lrcorner b$, $a \llcorner b$, $(a\ ,\ b)$ and, $a \cdot b$ respectively, without any need for prior conjugation.

At time of print the REDUCE package can be obtained either as source or precompiled for computers running Windows or Linux operating systems from the website http://sourceforge.net/projects/reduce-algebra/files/.

Algebraic Rules

```
%------------------------------------------------------------------------
% algebraic rules
%------------------------------------------------------------------------
algebraic;
operator e; noncom e; factor e;
clifrule:={e(~j)^2=>-1, e(~j)*e(~k)=>-e(k)*e(j) when ordp(j,k) and j neq k};
let clifrule;
%------------------------------------------------------------------------
```


A. Seagar, *Application of Geometric Algebra to Electromagnetic Scattering*,
DOI 10.1007/978-981-10-0089-8

Bit Operations

```
%------------------------------------------------------------------------
% bit operations
%------------------------------------------------------------------------
procedure andb(a,b);            % and least significant bits
if a=1 and b=1 then 1 else 0;

procedure orb(a,b);             % or least significant bits
if a=1 or b=1 then 1 else 0;

procedure lsb(x);               % get least significant bit
if evenp(x) then 0 else 1;

procedure rshift1(x);           % shift bits right, lose lsb
fix(x/2);

procedure nintersect(a,b);      % count number of bits in intersection
if a>0 and b>0
then nintersect(rshift1(a),rshift1(b))+andb(lsb(a),lsb(b)) else 0;
%------------------------------------------------------------------------
```

List Operations

```
%------------------------------------------------------------------------
% list operations
%------------------------------------------------------------------------
procedure tlist(x);             % test if x is a list
if arglength(x)>0 and part(x,0)=list then 1 else 0;

procedure dimc(x);              % get dimension of Clifford number
if tlist(x)=1 then 1+dimc(first(x)) else 0;

procedure zeroc(n);             % make Clifford zero of dimension n;
if n>0 then list(zeroc(n-1),zeroc(n-1)) else 0;

procedure gradec(x,k);          % get grade k of Clifford number
if tlist(x)=1
then if k>0 then list(gradec(first(x),k),gradec(second(x),k-1))
            else list(gradec(first(x),k),zeroc(dimc(second(x))))
else if k>0 then 0 else x;
%------------------------------------------------------------------------
procedure addls(x,s);           % add scalar s to all coefficients of
if tlist(x)=1                   %                     Clifford number
then list(addls(first(x),s),addls(second(x),s)) else x+s;

procedure multls(x,s);          % post multiply Clifford number x by
if tlist(x)=1                   %                              scalar s
then list(multls(first(x),s),multls(second(x),s)) else x*s;
%------------------------------------------------------------------------
```

Symbolic Operations

```
%------------------------------------------------------------------------
% symbolic operations
%------------------------------------------------------------------------
procedure makeindex(n);         % make Clifford index of dimension n
if n>0
then list(makeindex(n-1),addls(makeindex(n-1),2^(n-1))) else 0;
```

```
procedure makecliffvar(n,x,i);  % make Clifford variable of dimension n
if n>0
then list(makecliffvar(n-1,x,first(i)),makecliffvar(n-1,x,second(i)))
else mkid(x,i);

procedure makecliff(n,x);        % make Clifford variable of dimension n
makecliffvar(n,x,makeindex(n)); % units are implicit

procedure makeunits(n);          % make Clifford units of dimension n
if n>0
then list(makeunits(n-1),multls(makeunits(n-1),e(n-1))) else 1;

procedure printclf(x,u);         % print Clifford number in algebraic form
if tlist(x)=1
then printclf(first(x),first(u))+printclf(second(x),second(u)) else x*u;

procedure printcliff(x);         % print Clifford number in algebraic form
printclf(x,makeunits(dimc(x)));
%---------------------------------------------------------------------------
```

Inversion of Signs

```
%---------------------------------------------------------------------------
% inversion of signs
%---------------------------------------------------------------------------
procedure negc(x);              % negative of Clifford number
if tlist(x)=1
then list(negc(first(x)),negc(second(x))) else -x;

procedure invc(x);              % involution of Clifford number
if tlist(x)=1
then list(invc(first(x)),ninvc(second(x))) else x;

procedure ninvc(x);             % negative of involution of Clifford
if tlist(x)=1                   %                                   number
then list(ninvc(first(x)),invc(second(x))) else -x;

procedure revc(x);              % reversion of Clifford number
if tlist(x)=1
then list(revc(first(x)),cnjc(second(x))) else x;

procedure nrevc(x);             % negative of reversion of Clifford number
if tlist(x)=1
then list(nrevc(first(x)),ncnjc(second(x))) else -x;

procedure cnjc(x);              % conjugate of Clifford number
if tlist(x)=1
then list(cnjc(first(x)),nrevc(second(x))) else x;

procedure ncnjc(x);             % negative of conjugate of Clifford number
if tlist(x)=1
then list(ncnjc(first(x)),revc(second(x))) else -x;
%---------------------------------------------------------------------------
```

Addition and Subtraction

```
%---------------------------------------------------------------------------
% addition
%---------------------------------------------------------------------------
```

```
procedure addx(a,b);            % add two equal dimension Clifford numbers
if tlist(a)=1 and tlist(b)=1
then list(addx(first(a),first(b)),addx(second(a),second(b)))
else if tlist(a)=0 and tlist(b)=0
     then a+b else write "bad arguments to addx";

procedure add2(a,b,na,nb);      % reduce to numbers of equal dimension
if na=nb then addx(a,b)
else if na>nb
     then list(add2(first(a),b,na-1,nb),second(a))
     else list(add2(a,first(b),na,nb-1),second(b));

procedure addc(a,b);            % add two Clifford numbers
add2(a,b,dimc(a),dimc(b));
%----------------------------------------------------------------------
% subtraction
%----------------------------------------------------------------------
procedure subx(a,b);            % subtract two equal dimension
if tlist(a)=1 and tlist(b)=1    %                        Clifford numbers
then list(subx(first(a),first(b)),subx(second(a),second(b)))
else if tlist(a)=0 and tlist(b)=0
     then a-b else write "bad arguments to subx";

procedure sub2(a,b,na,nb);      % reduce to numbers of equal dimension
if na=nb then subx(a,b)
else if na>nb
     then list(sub2(first(a),b,na-1,nb),second(a))
     else list(sub2(a,first(b),na,nb-1),negc(second(b)));

procedure subc(a,b);            % subtract two Clifford numbers
sub2(a,b,dimc(a),dimc(b));
%----------------------------------------------------------------------
```

Products

```
%----------------------------------------------------------------------
% central product
%----------------------------------------------------------------------
procedure cmulx(a,b);           % central product for two equal dimension
if tlist(a)=1 and tlist(b)=1    %                        Clifford numbers
then list(subx(cmulx(first(a),first(b)),cmulx(second(a),invc(second(b)))),
          addx(cmulx(first(a),second(b)),cmulx(second(a),invc(first(b)))))
else if tlist(a)=0 and tlist(b)=0
     then a*b else write "bad arguments to cmulx";

procedure cmul2(a,b,na,nb);     % reduce to numbers of equal dimension
if na=nb then cmulx(a,b)
else if na>nb
     then list(cmul2(first(a),b,na-1,nb),
               cmul2(second(a),invc(b),na-1,nb))
     else list(cmul2(a,first(b),na,nb-1),
               cmul2(a,second(b),na,nb-1));

procedure central(a,b);         % central product of two Clifford numbers
cmul2(a,b,dimc(a),dimc(b));
%----------------------------------------------------------------------
% outer product
%----------------------------------------------------------------------
procedure omulx(a,b,n,i,j);     % outer product for two equal dimension
if tlist(a)=1 and tlist(b)=1    %                        Clifford numbers
then list(subx(omulx(first(a),first(b),n-1,i,j),
               omulx(second(a),invc(second(b)),n-1,2^(n-1)+i,2^(n-1)+j)),
          addx(omulx(first(a),second(b),n-1,i,2^(n-1)+j),
               omulx(second(a),invc(first(b)),n-1,2^(n-1)+i,j)))
else if tlist(a)=0 and tlist(b)=0
     then if nintersect(i,j)=0
```

```
          then a*b else 0
     else write "bad arguments to omulx";

procedure omul2(a,b,na,nb,i,j); % reduce to numbers of equal dimension
if na=nb then omulx(a,b,na,i,j)
else if na>nb
     then list(omul2(first(a),b,na-1,nb,i,j),
               omul2(second(a),invc(b),na-1,nb,2^(na-1)+i,j))
     else list(omul2(a,first(b),na,nb-1,i,j),
               omul2(a,second(b),na,nb-1,i,2^(nb-1)+j));

procedure outer(a,b);           % outer product of two Clifford numbers
omul2(a,b,dimc(a),dimc(b),0,0);
%-----------------------------------------------------------------------
% inner product
%-----------------------------------------------------------------------
procedure imulx(a,b,n,i,j);     % inner product for two equal dimension
if tlist(a)=1 and tlist(b)=1    %                     Clifford numbers
then list(subx(imulx(first(a),first(b),n-1,i,j),
               imulx(second(a),invc(second(b)),n-1,2^(n-1)+i,2^(n-1)+j)),
          addx(imulx(first(a),second(b),n-1,i,2^(n-1)+j),
               imulx(second(a),invc(first(b)),n-1,2^(n-1)+i,j)))
else if tlist(a)=0 and tlist(b)=0
     then if nintersect(i,j)=nintersect(i,i) or nintersect(i,j)=nintersect(j,j)
          then a*b else 0
     else write "bad arguments to imulx";

procedure imul2(a,b,na,nb,i,j); % reduce to numbers of equal dimension
if na=nb then imulx(a,b,na,i,j)
else if na>nb
     then list(imul2(first(a),b,na-1,nb,i,j),
               imul2(second(a),invc(b),na-1,nb,2^(na-1)+i,j))
     else list(imul2(a,first(b),na,nb-1,i,j),
               imul2(a,second(b),na,nb-1,i,2^(nb-1)+j));

procedure inner(a,b);           % inner product of two Clifford numbers
imul2(cnjc(a),b,dimc(a),dimc(b),0,0);
%-----------------------------------------------------------------------
% left inner product
%-----------------------------------------------------------------------
procedure lmulx(a,b,n,i,j);     % left inner product for two equal
if tlist(a)=1 and tlist(b)=1    %             dimension Clifford numbers
then list(subx(lmulx(first(a),first(b),n-1,i,j),
               lmulx(second(a),invc(second(b)),n-1,2^(n-1)+i,2^(n-1)+j)),
          addx(lmulx(first(a),second(b),n-1,i,2^(n-1)+j),
               lmulx(second(a),invc(first(b)),n-1,2^(n-1)+i,j)))
else if tlist(a)=0 and tlist(b)=0
     then if nintersect(i,i)=nintersect(i,j)
          then a*b else 0
     else write "bad arguments to lmulx";

procedure lmul2(a,b,na,nb,i,j); % reduce to numbers of equal dimension
if na=nb then lmulx(a,b,na,i,j)
else if na>nb
     then list(lmul2(first(a),b,na-1,nb,i,j),
               lmul2(second(a),invc(b),na-1,nb,2^(na-1)+i,j))
     else list(lmul2(a,first(b),na,nb-1,i,j),
               lmul2(a,second(b) na,nb-1,i,2^(nb-1)+j));

procedure left(a,b);            % left inner product of two Clifford
lmul2(cnjc(a),b,dimc(a),dimc(b),0,0);       %                   numbers



%-----------------------------------------------------------------------
% right inner product
%-----------------------------------------------------------------------
procedure rmulx(a,b,n,i,j);     % right inner product for two equal
if tlist(a)=1 and tlist(b)=1    %             dimension Clifford numbers
then list(subx(rmulx(first(a),first(b),n-1,i,j),
               rmulx(second(a),invc(second(b)),n-1,2^(n-1)+i,2^(n-1)+j)),
```

```
          addx(rmulx(first(a),second(b),n-1,i,2^(n-1)+j),
               rmulx(second(a),invc(first(b)),n-1,2^(n-1)+i,j)))
else if tlist(a)=0 and tlist(b)=0
     then if nintersect(i,j)=nintersect(j,j)
          then a*b else 0
     else write "bad arguments to rmulx";

procedure rmul2(a,b,na,nb,i,j); % reduce to numbers of equal dimension
if na=nb then rmulx(a,b,na,i,j)
else if na>nb
     then list(rmul2(first(a),b,na-1,nb,i,j),
               rmul2(second(a),invc(b),na-1,nb,2^(na-1)+i,j))
     else list(rmul2(a,first(b),na,nb-1,i,j),
               rmul2(a,second(b),na,nb-1,i,2^(nb-1)+j));

procedure right(a,b);            % right inner product of two Clifford
rmul2(a,cnjc(b),dimc(a),dimc(b),0,0);        %                    numbers
%---------------------------------------------------------------------------
% scalar product
%---------------------------------------------------------------------------
procedure smulx(a,b,n,i,j);      % scalar product for two equal dimension
if tlist(a)=1 and tlist(b)=1     %                      Clifford numbers
then list(subx(smulx(first(a),first(b),n-1,i,j),
               smulx(second(a),invc(second(b)),n-1,2^(n-1)+i,2^(n-1)+j)),
          addx(smulx(first(a),second(b),n-1,i,2^(n-1)+j),
               smulx(second(a),invc(first(b)),n-1,2^(n-1)+i,j)))
else if tlist(a)=0 and tlist(b)=0
     then if i=j
          then a*b else 0
     else write "bad arguments to smulx";

procedure smul2(a,b,na,nb,i,j); % reduce to numbers of equal dimension
if na=nb then smulx(a,b,na,i,j)
else if na>nb
     then list(smul2(first(a),b,na-1,nb,i,j),
               smul2(second(a),invc(b),na-1,nb,2^(na-1)+i,j))
     else list(smul2(a,first(b),na,nb-1,i,j),
               smul2(a,second(b),na,nb-1,i,2^(nb-1)+j));

% reserved word "scalar" cannot be used for name of procedure
procedure skalar(a,b);           % scalar product of two Clifford numbers
smul2(cnjc(a),b,dimc(a),dimc(b),0,0);
%---------------------------------------------------------------------------
% dot product
%---------------------------------------------------------------------------
procedure dmulx(a,b,n,i,j);      % dot product for two equal dimension
if tlist(a)=1 and tlist(b)=1     %                      Clifford numbers
then list(subx(dmulx(first(a),first(b),n-1,i,j),
               dmulx(second(a),invc(second(b)),n-1,2^(n-1)+i,2^(n-1)+j)),
          addx(dmulx(first(a),second(b),n-1,i,2^(n-1)+j),
               dmulx(second(a),invc(first(b)),n-1,2^(n-1)+i,j)))
else if tlist(a)=0 and tlist(b)=0
     then if i=j and nintersect(i,i)=1
          then a*b else 0
     else write "bad arguments to dmulx";

procedure dmul2(a,b,na,nb,i,j); % reduce to numbers of equal dimension
if na=nb then dmulx(a,b,na,i,j)
else if na>nb
     then list(dmul2(first(a),b,na-1,nb,i,j),
               dmul2(second(a),invc(b),na-1,nb,2^(na-1)+i,j))
     else list(dmul2(a,first(b),na,nb-1,i,j),
               dmul2(a,second(b),na,nb-1,i,2^(nb-1)+j));

procedure dot(a,b);              % dot product of two Clifford numbers
dmul2(cnjc(a),b,dimc(a),dimc(b),0,0);
%---------------------------------------------------------------------------
end;
%---------------------------------------------------------------------------
```

Appendix B
Electrical Properties of Materials

Table B.1 lists the electrical properties of a selection of materials of different types. They serve to illustrate the range of values these properties might well be expected to cover. The values are quoted to the number of decimal places as appearing in the sources.

The permittivity ϵ relates to the quadrature phase component of voltage and current (electric field and magnetic field) arising from bound charge (in the valence band). The loss tangent $\tan\delta$ relates to the in-phase component of voltage and current for bound charge in combination with the in-phase component of voltage and current for free charge (in the conduction band). The conductivity σ relates to the in-phase component of voltage and current for free charge alone. The conductivity and the loss tangent both play the role of loss mechanisms. For good conductors the effects of the free charge is dominant. For good insulators loss the effects of the bound charge dominate.

The three quantities are treated as constant within narrow ranges of frequency and temperature, but may take different values for markedly different frequencies or temperatures.

The permittivity covers a range of values as different atomic mechanisms become more or less significant as the frequency varies. The values listed are measured at microwave frequencies or lower, and are typically the highest values observed for the particular materials.

Conductivity varies with temperature, particularly in the case of semiconductors. The values listed are nominally at or close to room temperature. Conductivity is also sensitive to the purity of the materials, again in the case of semiconductors and also in the case of liquids (e.g. water) and crystalline solids (e.g. quartz). Naturally occurring samples are likely to have higher conductivity, often significantly so.

A. Seagar, *Application of Geometric Algebra to Electromagnetic Scattering*,
DOI 10.1007/978-981-10-0089-8

Table B.1 Electrical properties of materials of different types

Type	Material	Formula	Permittivity ϵ/ϵ_0	Loss tangent $\tan\delta$	Conductivity $\sigma/\Omega^{-1}m^{-1}$	Reference
	Air		1.0006	0	0	
Metal	Silver	Ag			6.80×10^7	[71]
	Copper	Cu			5.84×10^7	[72]
	Gold	Au			4.5×10^7	[73]
	Aluminium	Al			3.69×10^7	[72]
	Platinum	Pt			9.43×10^6	[71]
	Mercury	Hg			1.06×10^6	[72]
Semi-metal	Antimony	Sb			2.56×10^6	[72]
	Arsenic	As			2.7×10^6	[73]
	Graphite	C			1.82×10^6	[74]
	Bismuth	Bi			9.35×10^5	[72]
Semiconductor	Germanium	Ge	16.3		2.23	[75]
	Silicon	Si	11.7		4.25×10^{-4}	[75]
	Gallium arsenide	GaAs	12		1.27×10^{-6}	[75]
	Silicon carbide	SiC	10.2			[72]
H-C-O	Sea water	H_2O + salts	81		5.3088	[29, 72]
	Acetone	$(CH_3)_2CO$	21.01		6×10^{-6}	[72, 76]
	Water	H_2O	80.2		5.5×10^{-6}	[72, 73]
	Acetic acid	CH_3CO_2H	6.20		1.12×10^{-6}	[72, 76]
	Ethanol	C_2H_6O	25.3		1.35×10^{-7}	[72, 76]
	Octane	C_8H_{18}	1.948			[72]

(continued)

Table B.1 (continued)

Type	Material	Formula	Permittivity ϵ/ϵ_0	Loss tangent $\tan\delta$	Conductivity $\sigma/\Omega^{-1}\text{m}^{-1}$	Reference
	Air		1.0006	0	0	
Glass	Silica	SiO_2	3.81	0.00038	2.0×10^{-8}	[71, 72]
	Sodium borosilicate (pyrex)		5.07	0.019		[72]
	Lead oxide (crystal)		6.9			[77]
	Soda-lime (window)		7.5			[77]
Ceramic	Sapphire	Al_2O_3	11.5	0.0001	1×10^{-9}	[71]
	Quartz (natural)	SiO_2	4.60		1.0×10^{-10}	[72, 78]
	Diamond	C	5.87	0.00020	1×10^{-11}	[71–73]
	Quartz (synthetic)	SiO_2	4.60		1.2×10^{-17}	[72, 78]
	Lead zirconate titanate (PZT)	$PbZrO_3 + PbTiO_3$	1800	0.015		[71]
Polymer	Polyvinylchloride (PVC)		3.0	0.015	1.0×10^{-13}	[71]
	Polyethylenetere-phthalate (PET)		3.19	0.00852	2.79×10^{-15}	[71]
	Polycarbonate		2.97	0.00551	2.21×10^{-15}	[71]
	Polypropylene		2.3	0.000410	8.13×10^{-16}	[71]
	Polystyrene		2.58		4.41×10^{-16}	[71]
	Polyethylene		2.3			[77]

Appendix C
C Code for One-Dimensional Solution

The source code listed here is a verbatim copy of that used to create the results given in Fig. 10.3. Prior to compilation a header file should be created using the C programming utility *cproto* or equivalent, as shown in the first line of Table C.1. The third line will generate a usage message with examples.

The fourth line in Table C.1 is an example of usage with use of the file operator '>' on the UNIX command line to redirect the ASCII output into a plain text file *output.txt*. In this case the material is gold, with material properties from left to right: relative permeability $\mu/\mu_0 = 1$, relative permittivity $\epsilon/\epsilon_0 = 1$, loss tangent $\tan\delta = 0$, electrical conductivity $\sigma = 4.5 \times 10^7$ Siemens per metre, thickness $x = 10$ free space wavelengths, frequency $f = 2 \times 10^9$ Hz. The text file produced is suitable for plotting directly with a graphics utility such as *gnuplot*.

The parameters on the command line provide two methods for specifying the lossy component of a material's properties; either as $\tan\delta$, the loss tangent, or as σ, the electrical conductivity. Within the program the losses are added together. When only one is required, a value of zero can be used for the other. For the results in Fig. 10.3 both, where available, have been used because the tests are intended to examine behaviour for the most extreme values which may be expected.

```
/*---------------------------------------------------------*/
/* one.c
   Implementation of Cauchy-Clifford-Dirac technique
   in one dimension.                                        */
/*---------------------------------------------------------*/
/* structure for electromagnetic field                      */
/*---------------------------------------------------------*/
/* f = A.e3.sigma+i.B.e2.e0   */
/*     A = +sqrt(mu)*Hz       */
/*     B = -sqrt(epsilon)*Ey  */
typedef struct xxx{
  double A[2]; /* complex value of +sqrt(mu)*Hz       */
  double B[2]; /* complex value of -sqrt(epsilon)*Ey  */
}field;        /* structure for electromagnetic field */
/*---------------------------------------------------------*/
#include <math.h>
#include <stdio.h>
#include <stdlib.h>
#include "one.h"  /* header file with function prototypes */
/*---------------------------------------------------------*/
```


A. Seagar, *Application of Geometric Algebra to Electromagnetic Scattering*,
DOI 10.1007/978-981-10-0089-8

Table C.1 Compilation and usage of C code

$ cproto one.c > one.h
$ cc –c one one.c –lm
$ one
$ one 1 1 0 4.5e7 10 2e9 > output.txt

```
/* real-valued routines */
/*-------------------------------------------------------*/
double velocity(omega,k)
     double omega;
     double k[2];
{
  double v;

  v=omega/k[0];
  return(v);
}
/*-------------------------------------------------------*/
double wavelength(k)
     double k[2];
{
  double lambda;

  lambda=2.0*M_PI/k[0];
  return(lambda);
}
/*-------------------------------------------------------*/
double gain_dB(k)
     double k[2];
{
  double g;

  g=20.0*k[1]/(log(10.0));
  return(g);
}
/*-------------------------------------------------------*/
double contrast(k_minus,k_plus)
     double k_minus[2],k_plus[2];
{
  double r[2];
  double g;

  complex_divide(k_plus,k_minus,r);
  g=complex_magnitude(r);
  return(g);
}
/*-------------------------------------------------------*/
/* complex-valued routines */
/*-------------------------------------------------------*/
void complex_negate(x)
     double x[2]; /* complex */
{
  x[0]=-x[0];
  x[1]=-x[1];
}
/*-------------------------------------------------------*/
double complex_magnitude(x)
     double x[2]; /* complex */
{
  double r;

  r=sqrt(x[0]*x[0]+x[1]*x[1]);
  return(r);
}
/*-------------------------------------------------------*/
void complex_add(a,b,x)
```

```
     double a[2],b[2],x[2]; /* complex */
{
  x[0]=a[0]+b[0];
  x[1]=a[1]+b[1];
}
/*------------------------------------------------------*/
void complex_subtract(a,b,x)
     double a[2],b[2],x[2]; /* complex */
{
  x[0]=a[0]-b[0];
  x[1]=a[1]-b[1];
}

/*------------------------------------------------------*/
void complex_multiply(a,b,x)
     double a[2],b[2],x[2]; /* complex */
{
  x[0]=a[0]*b[0]-a[1]*b[1];
  x[1]=a[1]*b[0]+a[0]*b[1];
}
/*------------------------------------------------------*/
void complex_divide(a,b,x)
     double a[2],b[2],x[2]; /* complex */
{
  double r;

  r=1.0/(b[0]*b[0]+b[1]*b[1]);
  x[0]=r*(a[0]*b[0]+a[1]*b[1]);
  x[1]=r*(a[1]*b[0]-a[0]*b[1]);
}
/*------------------------------------------------------*/
void complex_sqrt(a,x)
     double a[2],x[2]; /* complex */
{
  double r,theta;

  if(a[1]==0.0) /* real */
    {
      if(a[0]<0.0) /* negative */
        {
          x[0]=0.0;
          x[1]=sqrt(-a[0]);
        }
      else /* non-negative */
        {
          x[0]=sqrt(a[0]);
          x[1]=0.0;
        }
    }
  else /* non-real */
    {
      r=sqrt(a[0]*a[0]+a[1]*a[1]);
      if(a[1]>0.0) /* above real axis */
        {
          theta=0.5*atan2(a[1],a[0]);
        }
      else /* not above real axis */
        {
          theta=-0.5*atan2(-a[1],a[0]);
        }
      x[0]=r*cos(theta);
      x[1]=r*sin(theta);
    }
}
/*------------------------------------------------------*/
void emikx(k,x,e)      /* e^(-ikx) */
     double k[2],e[2]; /* complex */
     double x;
{
  double r,theta;
```

```
  theta=k[0]*x;
  r=exp(k[1]*x);
  e[0]= r*cos(theta);
  e[1]=-r*sin(theta);
}
/*--------------------------------------------------------*/
void wavenumber(omega,sigma,mu,epsilon,k)
     double omega,sigma,mu;    /* real */
     double epsilon[2],k[2];   /* complex */
{
  double x[2];
  double wu;

  wu=omega*mu;
  x[0]=wu*omega*epsilon[0];
  x[1]=wu*(omega*epsilon[1]-sigma);
  complex_sqrt(x,k);
}
/*--------------------------------------------------------*/
void space_factor(mu_plus,mu_minus,s_scale)
     double mu_minus,mu_plus; /* real */
     double s_scale[2];       /* complex */
{
  double x[2],y[2],a[2],b[2];

  x[0]=mu_plus;
  x[1]=0.0;
  y[0]=mu_minus;
  y[1]=0.0;
  complex_sqrt(x,a);
  complex_sqrt(y,b);
  complex_subtract(a,b,x);
  complex_add(a,b,y);
  complex_divide(x,y,s_scale);
}
/*--------------------------------------------------------*/
void time_factor(omega,sigma_plus,sigma_minus,epsilon_plus,epsilon_minus,t_scale)
     double omega,sigma_minus,sigma_plus;      /* real */
     double epsilon_minus[2],epsilon_plus[2]; /* complex */
     double t_scale[2];                        /* complex */
{
  double x[2],y[2],a[2],b[2];

  x[0]=omega*epsilon_minus[0];
  x[1]=omega*epsilon_minus[1]-sigma_minus;
  y[0]=omega*epsilon_plus[0];
  y[1]=omega*epsilon_plus[1]-sigma_plus;
  complex_sqrt(x,a);
  complex_sqrt(y,b);
  complex_subtract(a,b,x);
  complex_add(a,b,y);
  complex_divide(x,y,t_scale);
}
/*--------------------------------------------------------*/
/* field conversion routines */
/* f = A.e3.sigma+i.B.e2.e0   */
/*     A = +sqrt(mu)*Hz       */
/*     B = -sqrt(epsilon)*Ey */
/*--------------------------------------------------------*/
void set_field(Hz,Ey,omega,sigma,mu,epsilon,f)
     double Hz[2],Ey[2],epsilon[2]; /* complex */
     double omega,sigma,mu;         /* real */
     field f[1];                    /* field at one point */
{
  double x[2],y[2],a[2],b[2];

  x[0]=mu;
  x[1]=0.0;
  y[0]=epsilon[0];
  y[1]=epsilon[1]-sigma/omega;
  complex_sqrt(x,a);
```

```
  complex_sqrt(y,b);
  complex_negate(b);
  complex_multiply(a,Hz,f[0].A);
  complex_multiply(b,Ey,f[0].B);
}
/*--------------------------------------------------------*/
void get_field(f,omega,sigma,mu,epsilon,Hz,Ey)
     field f[1];                      /* field at one point */
     double omega,sigma,mu,epsilon; /* real */
     double Hz[2],Ey[2];              /* complex */
{
  double x[2],y[2],a[2],b[2];

  x[0]= mu;
  x[1]= 0.0;
  y[0]= epsilon;
  y[1]=-sigma/omega;
  complex_sqrt(x,a);
  complex_sqrt(y,b);
  complex_negate(b);
  complex_divide(f[0].A,a,Hz);
  complex_divide(f[0].B,b,Ey);
}
/*--------------------------------------------------------*/
/* single point field-valued routines */
/*--------------------------------------------------------*/
void complex_scale_field(s,a,x)
     double s[2];      /* complex */
     field a[1],x[1]; /* field at one point */
{
  complex_multiply(s,a[0].A,x[0].A);
  complex_multiply(s,a[0].B,x[0].B);
}

/*--------------------------------------------------------*/
void one_plus_ie0e1(f,x) /* x=[1+ie(0)e(1)]f */
     field f[1],x[1];    /* field at one point */
{
  x[0].A[0] = f[0].A[0] + f[0].B[0];
  x[0].A[1] = f[0].A[1] + f[0].B[1];
  x[0].B[0] = f[0].B[0] + f[0].A[0];
  x[0].B[1] = f[0].B[1] + f[0].A[1];
}
/*--------------------------------------------------------*/
void one_minus_ie0e1(f,x) /* x=[1-ie(0)e(1)]f */
     field f[1],x[1];     /* field at one point */
{
  x[0].A[0] = f[0].A[0] - f[0].B[0];
  x[0].A[1] = f[0].A[1] - f[0].B[1];
  x[0].B[0] = f[0].B[0] - f[0].A[0];
  x[0].B[1] = f[0].B[1] - f[0].A[1];
}
/*--------------------------------------------------------*/
void half_add_field(a,b,x) /* x=(a+b)/2 */
     field a[1],b[1],x[1]; /* field at one point */
{
  x[0].A[0] = 0.5*(a[0].A[0] + b[0].A[0]);
  x[0].A[1] = 0.5*(a[0].A[1] + b[0].A[1]);
  x[0].B[0] = 0.5*(a[0].B[0] + b[0].B[0]);
  x[0].B[1] = 0.5*(a[0].B[1] + b[0].B[1]);
}
/*--------------------------------------------------------*/
void half_subtract_field(a,b,x) /* x=(a-b)/2 */
     field a[1],b[1],x[1];      /* field at one point */
{
  x[0].A[0] = 0.5*(a[0].A[0] - b[0].A[0]);
  x[0].A[1] = 0.5*(a[0].A[1] - b[0].A[1]);
  x[0].B[0] = 0.5*(a[0].B[0] - b[0].B[0]);
  x[0].B[1] = 0.5*(a[0].B[1] - b[0].B[1]);
}
/*--------------------------------------------------------*/
```

```
/* whole surface field-valued routines */
/*-------------------------------------------------------*/
void guess_zero(x) /* x=0 */
     field x[2];   /* field over whole surface */
{
  x[0].A[0] = 0.0;  x[0].A[1] = 0.0;
  x[0].B[0] = 0.0;  x[0].B[1] = 0.0;
  x[1].A[0] = 0.0;  x[1].A[1] = 0.0;
  x[1].B[0] = 0.0;  x[1].B[1] = 0.0;
}
/*-------------------------------------------------------*/
void guess_f_minus(x) /* x=guess */
     field x[2];      /* field over whole surface */
{
  guess_zero(x);
}
/*-------------------------------------------------------*/
void guess_f_plus(x) /* x=guess */
     field x[2];     /* field over whole surface */
{
  guess_zero(x);
}
/*-------------------------------------------------------*/
double rms_field(x) /* RMS field over whole surface */
     field x[2];    /* field over whole surface */
{
  double f;

  f = (x[0].A[0] * x[0].A[0]) + (x[0].A[1] * x[0].A[1])
    + (x[0].B[0] * x[0].B[0]) + (x[0].B[1] * x[0].B[1])
    + (x[1].A[0] * x[1].A[0]) + (x[1].A[1] * x[1].A[1])
    + (x[1].B[0] * x[1].B[0]) + (x[1].B[1] * x[1].B[1]);
  f=sqrt(0.5*f);
  return(f);
}
/*-------------------------------------------------------*/
void invert_whole_field(x) /* x=-x */
     field x[2];           /* field over whole surface */
{
  x[0].A[0] = -x[0].A[0];  x[0].A[1] = -x[0].A[1];
  x[0].B[0] = -x[0].B[0];  x[0].B[1] = -x[0].B[1];
  x[1].A[0] = -x[1].A[0];  x[1].A[1] = -x[1].A[1];
  x[1].B[0] = -x[1].B[0];  x[1].B[1] = -x[1].B[1];
}
/*-------------------------------------------------------*/
void copy_whole_field(a,x) /* x=a */
     field a[2],x[2];      /* field over whole surface */
{
  x[0].A[0] = a[0].A[0];  x[0].A[1] = a[0].A[1];
  x[0].B[0] = a[0].B[0];  x[0].B[1] = a[0].B[1];
  x[1].A[0] = a[1].A[0];  x[1].A[1] = a[1].A[1];
  x[1].B[0] = a[1].B[0];  x[1].B[1] = a[1].B[1];
}
/*-------------------------------------------------------*/
void add_whole_field(a,b,x) /* x=a+b */
     field a[2],b[2],x[2];  /* field over whole surface */
{
  x[0].A[0] = a[0].A[0] + b[0].A[0];
  x[0].A[1] = a[0].A[1] + b[0].A[1];
  x[0].B[0] = a[0].B[0] + b[0].B[0];
  x[0].B[1] = a[0].B[1] + b[0].B[1];
  x[1].A[0] = a[1].A[0] + b[1].A[0];
  x[1].A[1] = a[1].A[1] + b[1].A[1];
  x[1].B[0] = a[1].B[0] + b[1].B[0];
  x[1].B[1] = a[1].B[1] + b[1].B[1];
}
/*-------------------------------------------------------*/
void subtract_whole_field(a,b,x) /* x=a-b */
     field a[2],b[2],x[2]; /* field over whole surface */
{
  x[0].A[0] = a[0].A[0] - b[0].A[0];
```

```
  x[0].A[1] = a[0].A[1] - b[0].A[1];
  x[0].B[0] = a[0].B[0] - b[0].B[0];
  x[0].B[1] = a[0].B[1] - b[0].B[1];
  x[1].A[0] = a[1].A[0] - b[1].A[0];
  x[1].A[1] = a[1].A[1] - b[1].A[1];
  x[1].B[0] = a[1].B[0] - b[1].B[0];
  x[1].B[1] = a[1].B[1] - b[1].B[1];
}
/*--------------------------------------------------------*/
/* whole field-valued operators */
/*--------------------------------------------------------*/
void source_field(Jy,xa,xb,omega,sigma,mu,epsilon,f_star)
     double Jy[2],epsilon[2];      /* complex */
     double xa,xb,omega,sigma,mu; /* real*/
     field f_star[2];              /* field over whole surface */
{
  double x[2],y[2],a[2],b[2],e[2],k[2],Hz[2],Ey[2];

  x[0]= mu;
  x[1]= 0.0;
  y[0]= epsilon[0];
  y[1]= epsilon[1]-sigma/omega;
  complex_sqrt(x,a);
  complex_sqrt(y,b);
  a[0]=2.0*a[0];  a[1]=2.0*a[1];
  b[0]=2.0*b[0];  b[1]=2.0*b[1];
  complex_divide(Jy,a,x);
  complex_divide(Jy,b,y);
  wavenumber(omega,sigma,mu,epsilon,k);
  emikx(k,xa,e);
  complex_multiply(x,e,Hz);
  complex_multiply(y,e,Ey);
  set_field(Hz,Ey,omega,sigma,mu,epsilon,f_star);
  emikx(k,xb,e);
  complex_multiply(x,e,Hz);
  complex_multiply(y,e,Ey);
  set_field(Hz,Ey,omega,sigma,mu,epsilon,f_star+1);
}
/*--------------------------------------------------------*/
void RQ(f,s_scale,t_scale,x)        /* x=RQf */
     field f[2],x[2];               /* field over whole surface */
     double s_scale[2],t_scale[2]; /* complex */
{
  complex_multiply(s_scale,f[0].A,x[0].A);
  complex_multiply(t_scale,f[0].B,x[0].B);
  complex_multiply(s_scale,f[1].A,x[1].A);
  complex_multiply(t_scale,f[1].B,x[1].B);
}
/*--------------------------------------------------------*/
void half_I_plus_RQ(f,s_scale,t_scale,x) /* x=0.5*(I+RQ)f */
     field f[2],x[2];                    /* field over whole surface */
     double s_scale[2],t_scale[2];       /* complex */
{
  field RQf[2];

  RQ(f,s_scale,t_scale,RQf);
  half_add_field(f,RQf,x);
  half_add_field(f+1,RQf+1,x+1);
}
/*--------------------------------------------------------*/
void half_I_minus_RQ(f,s_scale,t_scale,x) /* x=0.5*(I-RQ)f */
     field f[2],x[2];                     /* field over whole surface */
     double s_scale[2],t_scale[2];        /* complex */
{
  field RQf[2];

  RQ(f,s_scale,t_scale,RQf);
  half_subtract_field(f,RQf,x);
  half_subtract_field(f+1,RQf+1,x+1);
}
/*--------------------------------------------------------*/
void boundary_data(f_star,s_scale,t_scale,g)  /* x=-0.5*(I+RQ)f_star */
     field f_star[2],g[2];                    /* field over whole surface */
```

```
     double s_scale[2],t_scale[2];            /* complex */
{
  field RQf[2];

  RQ(f_star,s_scale,t_scale,RQf);
  half_add_field(f_star,RQf,g);
  half_add_field(f_star+1,RQf+1,g+1);
  invert_whole_field(g);
}
/*-------------------------------------------------------*/
void Pp(f,k,bma,x)      /* x=P^(+)f */
     field f[2],x[2]; /* field over whole surface */
     double k[2];      /* complex */
     double bma;
{
  double e[2];
  field fa[1],fb[1],ef[1];

  emikx(k,bma,e);

  one_minus_ie0e1(f,fa);
  one_plus_ie0e1(f+1,fb);
  complex_scale_field(e,fb,ef);
  half_add_field(fa,ef,x);

  one_plus_ie0e1(f+1,fb);
  one_minus_ie0e1(f,fa);
  complex_scale_field(e,fa,ef);
  half_add_field(fb,ef,x+1);
}
/*-------------------------------------------------------*/
void Pm(f,k,bma,x)      /* x=P^(-)f */
     field f[2],x[2]; /* field over whole surface */
     double k[2];      /* complex */
     double bma;
{
  double e[2];
  field fa[1],fb[1],ef[1];

  emikx(k,bma,e);

  one_plus_ie0e1(f,fa);
  one_plus_ie0e1(f+1,fb);
  complex_scale_field(e,fb,ef);
  half_subtract_field(fa,ef,x);

  one_minus_ie0e1(f+1,fb);
  one_minus_ie0e1(f,fa);
  complex_scale_field(e,fa,ef);
  half_subtract_field(fb,ef,x+1);
}
/*-------------------------------------------------------*/
/* whole-surface field-valued iterations */
/*-------------------------------------------------------*/
void f2_plus(f1_minus,f0_plus,s_scale,t_scale,g,k_plus,bma,x)
     field f1_minus[2],f0_plus[2],g[2],x[2];
     double bma;
     double s_scale[2],t_scale[2],k_plus[2];
     /* x = P^(++)[0.5(I+RQ)f1^(-)-g+0.5(I+RQ)f0^(+)] */
{
  field x1[2],x2[2],x3[2],x4[2];

  half_I_plus_RQ(f1_minus,s_scale,t_scale,x1);
  half_I_plus_RQ(f0_plus,s_scale,t_scale,x3);
  subtract_whole_field(x1,g,x2);
  add_whole_field(x2,x3,x4);
  Pp(x4,k_plus,bma,x);
}
/*-------------------------------------------------------*/
void f3_minus(f2_plus,f1_minus,s_scale,t_scale,g,k_minus,bma,x)
     field f2_plus[2],f1_minus[2],g[2],x[2];
```

```
     double bma;
     double s_scale[2],t_scale[2],k_minus[2];
     /* x = P^(--)[0.5(I-RQ)f2^(+)+g+0.5(I-RQ)f1^(-)] */
{
  field x1[2],x2[2],x3[2],x4[2];

  half_I_minus_RQ(f2_plus,s_scale,t_scale,x1);
  half_I_minus_RQ(f1_minus,s_scale,t_scale,x3);
  add_whole_field(x1,g,x2);
  add_whole_field(x2,x3,x4);
  Pm(x4,k_minus,bma,x);
}
/*-------------------------------------------------------*/
/* single whole surface field-valued update */
/*-------------------------------------------------------*/
void update(f_minus,f_plus,s_scale,t_scale,g,k_minus,k_plus,bma,x_minus,x_plus)
     field f_minus[2],f_plus[2],g[2],x_minus[2],x_plus[2];
     double bma;
     double s_scale[2],t_scale[2],k_minus[2],k_plus[2];
{
  f2_plus(f_minus,f_plus,s_scale,t_scale,g,k_plus,bma,x_plus);
  f3_minus(x_plus,f_minus,s_scale,t_scale,g,k_minus,bma,x_minus);
}
/*-------------------------------------------------------*/
static double err_old;
/*-------------------------------------------------------*/
double error(n,f_minus,f_plus,s_scale,t_scale,g)
     int n;
     double s_scale[2],t_scale[2];
     field f_minus[2],f_plus[2],g[2];
{
  field g_approx[2],g_error[2],a_approx[2],b_approx[2];
  double err;

  half_I_plus_RQ(f_minus,s_scale,t_scale,b_approx);
  half_I_minus_RQ(f_plus,s_scale,t_scale,a_approx);
  subtract_whole_field(b_approx,a_approx,g_approx);
  subtract_whole_field(g_approx,g,g_error);
  err=rms_field(g_error)/rms_field(g);

  if(n==1) err_old=err;
  printf("%d %e %e %e\n",n,err,err/err_old,pow(err,1.0/n));
  err_old=err;

  return(err);
}
/*-------------------------------------------------------*/
/* multiple whole surface field-valued update */
/*-------------------------------------------------------*/
void iterate(n,s_scale,t_scale,g,k_minus,k_plus,bma,x_minus,x_plus)
     int n;
     field g[2],x_minus[2],x_plus[2];
     double bma;
     double s_scale[2],t_scale[2],k_minus[2],k_plus[2];
{
  field f_minus_old[2],f_plus_old[2],f_minus_new[2],f_plus_new[2];
  double err,old_err;
  int i;

  guess_f_minus(f_minus_old);
  guess_f_plus(f_plus_old);
  old_err=error(1,f_minus_old,f_plus_old,s_scale,t_scale,g);
  update(f_minus_old,f_plus_old,s_scale,t_scale,g,k_minus,k_plus,bma,f_minus_new,f_plus_new);
  err=error(2,f_minus_new,f_plus_new,s_scale,t_scale,g);
  for(i=3;i<=n;i++)
    {
      copy_whole_field(f_minus_new,f_minus_old);
      copy_whole_field(f_plus_new,f_plus_old);
      old_err=err;
      update(f_minus_old,f_plus_old,s_scale,t_scale,g,k_minus,k_plus,bma,f_minus_new,f_plus_new);
      err=error(i,f_minus_new,f_plus_new,s_scale,t_scale,g);
    }
```

```
  copy_whole_field(f_minus_new,x_minus);
  copy_whole_field(f_plus_new,x_plus);
  printf("# %d error = %e, rate of convergence = %e\n",i-1,err,err/old_err);
  printf("# dissipation, imag(k)/real(k) = %e\n",k_plus[1]/k_plus[0]);
  printf("### %e %e %e %e\n",err/old_err,gain_dB(k_plus),contrast(k_minus,k_plus),k_plus[1]);
  printf("\n\n");
}
/*-------------------------------------------------------*/
/* whole solution */
/*-------------------------------------------------------*/
void solve(a,b,omega,mu_minus,epsilon_minus,sigma_minus,mu_plus,epsilon_plus,sigma_plus)
     double a,b;                            /* real */
     double omega;                          /* real */
     double mu_minus,mu_plus;               /* real */
     double epsilon_minus[2],epsilon_plus[2]; /* complex */
     double sigma_minus,sigma_plus;         /* real */
{
  double k_minus[2],k_plus[2];       /* complex */
  double t_scale[2],s_scale[2];      /* complex */
  double Jy[2];                      /* complex */
  field g[2],f_star[2],f_minus[2],f_plus[2];   /* fields over whole surface */
  double bma,lambda,v,gdB;
  int n;

  bma=b-a;
  wavenumber(omega,sigma_minus,mu_minus,epsilon_minus,k_minus);
  wavenumber(omega,sigma_plus,mu_plus,epsilon_plus,k_plus);

  lambda=wavelength(k_minus);
  v=velocity(omega,k_minus);
  gdB=gain_dB(k_minus);
  printf("# region (-): wavelength = %e m, velocity = %e m/s, gain = %e dB/m.\n",lambda,v,gdB);

  lambda=wavelength(k_plus);
  v=velocity(omega,k_plus);
  gdB=gain_dB(k_plus);
  printf("# region (+): wavelength = %e m, velocity = %e m/s, gain = %e dB/m.\n",lambda,v,gdB);

  space_factor(mu_plus,mu_minus,s_scale);
  time_factor(omega,sigma_plus,sigma_minus,epsilon_plus,epsilon_minus,t_scale);

  Jy[0]=1.0;
  Jy[1]=0.0;
  source_field(Jy,a,b,omega,sigma_minus,mu_minus,epsilon_minus,f_star);
  boundary_data(f_star,s_scale,t_scale,g);

  n=50;
  iterate(n,s_scale,t_scale,g,k_minus,k_plus,bma,f_minus,f_plus);
}
/*-------------------------------------------------------*/
#define MU0  (4.0e-7*M_PI)
#define C0   (299792458.0)
#define EPS0 (1.0/(MU0*C0*C0))
/*-------------------------------------------------------*/
void get_specifications(argc,argv,mu_plus,epsilon_plus,sigma_plus,b,omega)
     int argc; char **argv;
     double *mu_plus,epsilon_plus[2],*sigma_plus,*b,*omega;
{
     double mu_mu0,eps_eps0,sigma,b_lambda0,f,tan_delta;
     int i;

     if(argc!=7)
       {
         printf("found   %d arguments:",argc-1);
         for(i=1;i<argc;i++)
           {
             printf(" %s",argv[i]);
           }
         printf("\nrequire 6 arguments: ");
         printf("mu/mu_0 epsilon/epsilon_0 tan(delta) sigma thickness/lambda_0 frequency\n");
         printf("example (gold): one  1  1     0      4.5e7   10  2e9\n");
         printf("     (diamond): one  1  5.87  0.0002  1e-11   10  2e9\n");
```

```
        printf("   (sea water): one  1  81    0      5.3088  10  2e9\n");
        exit(0);
      }

    sscanf(argv[1],"%lf",&mu_mu0);
    sscanf(argv[2],"%lf",&eps_eps0);
    sscanf(argv[3],"%lf",&tan_delta);
    sscanf(argv[4],"%lf",&sigma);
    sscanf(argv[5],"%lf",&b_lambda0);
    sscanf(argv[6],"%lf",&f);

    (*mu_plus)=mu_mu0*MU0;
    epsilon_plus[0]=eps_eps0*EPS0;
    epsilon_plus[1]=-tan_delta*epsilon_plus[0];
    (*sigma_plus)=sigma;
    (*b)=b_lambda0*C0/f;
    (*omega)=2.0*M_PI*f;
}
/*--------------------------------------------------------*/

int main(argc,argv)
    int argc; char **argv;
{
  double a,b,f;                                /* real */
  double omega;                                /* real */
  double mu_minus,mu_plus;                     /* real */
  double epsilon_minus[2],epsilon_plus[2];     /* complex */
  double sigma_minus,sigma_plus;               /* real */
  int i;

  a=0.0;                 /* metres             */
  mu_minus=MU0;          /* Henries per metre  */
  epsilon_minus[0]=EPS0; /* Farads per metre   */
  epsilon_minus[1]=0.0;
  sigma_minus=0.0;       /* 1/(Ohms) per metre */
  get_specifications(argc,argv,&mu_plus,epsilon_plus,&sigma_plus,&b,&omega);

  f=omega/(2.0*M_PI);
  printf("# frequency = %e Hz.\n",f);
  printf("# region (-): rel. permeability %f\n",mu_minus/MU0);
  printf("# region (-): rel. permittivity %f, loss tangent %e (bound+free charge).\n"
         epsilon_minus[0]/EPS0,-epsilon_minus[1]/epsilon_minus[0]);
  printf("# region (-): conductivity %e (1/ohm)/m, loss tangent %e (free charge).\n",
         sigma_minus,sigma_minus/(omega*epsilon_minus[0]));
  printf("# region (+): thickness %e m, %e wavelengths.\n",b,b*f/C0);
  printf("# region (+): rel. permeability %f.\n",mu_plus/MU0);
  printf("# region (+): rel. permittivity %f, loss tangent %e (bound-free charge).\n".
         epsilon_plus[0]/EPS0,-epsilon_plus[1]/epsilon_plus[0]);
  printf("# region (+): conductivity %e (1/ohm)/m, loss tangent %e (free charge).\n",
         sigma_plus,sigma_plus/(omega*epsilon_plus[0]));

  solve(a,b,omega,mu_minus,epsilon_minus,sigma_minus,mu_plus,epsilon_plus,sigma_plus);

  return(0);
}
/*--------------------------------------------------------*/
```

References

1. Maxwell JC (1873) A treatise on electricity and magnetism, vol 2. Clarendon Press, Oxford
2. Axelsson A, Grognard R, Hogan J, McIntosh A (2001) Harmonic analysis of Dirac operators on Lipschitz domains. In: Brackx F (ed) Clifford analysis and its applications. NATO science series II, mathematics physics and chemistry, vol 25. Kluwer, Dordrecht, pp 231–246
3. Clifford WK (1878) Applications of Grassmann's extensive algebra. Am J Math 1(4):350–358
4. Chisholm M (2002) Such silver currents: the story of William and Lucy Clifford 1845–1929. Lutterworth Press, Cambridge
5. Sir William RH (1844) On quaternions; or on a new system of imaginaries in algebra. Philos Mag J Sci 25:10–13
6. Grassmann H (1995) The extension theory of 1844. A new branch of mathematics: the ausdehnungslehre of 1844, and other works. Open Court, Chicago, pp 2–295 (translated by Lloyd C. Kannenberg)
7. Grassmann H (2000) Extension theory. History of mathematics, vol 19. American Mathematical Society, Providence
8. Cauchy BA (1841) Résumé d'un mémoire sur la mécanique céleste et sur un nouveau calcul appelé calcul des limites. In: Bachelier (ed) Exercices d'Analyse et de Physique Mathématique, vol 2. Bachelier, Paris, pp 41–98
9. Smithies F (1997) Cauchy and the creation of complex function theory. Cambridge University Press, Cambridge
10. McIntosh A (1989) Clifford algebras and the higher-dimensional Cauchy integral. In: Ciesielski Z (ed) Approximation and function spaces., number 22 in BCP seriesBanach Center Publications, Warsaw, pp 253–267
11. McIntosh A (1996) Clifford algebras, Fourier theory, singular integrals and harmonic functions on Lipschitz domains. In: Ryan J (ed) Clifford algebras in analysis and related topics. Studies in advanced mathematics, vol 21. CRC Press, Boca Raton, pp 33–87
12. Andraes A (2002) Transmission problems for Dirac's and Maxwell's equations with Lipschitz Interfaces. Ph.D. thesis, Australian National University, Canberra, Australia
13. Wilson EB (1901) Vector analysis: a text-book for the use of students of mathematics and physics. Charles Scribner's Sons, New York
14. Heaviside O (1893) Electromagnetic theory, vol 1. The Electrician Printing and Publishing Company, London
15. Gill F (1925) Oliver Heaviside. Bell Syst Tech J 4:349–354
16. Unz H (1963) Oliver heaviside (1850–1925). IEEE Trans Educ 6(1):30–33
17. Grassmann H (1855) Sur les différents genres de multiplication. Journal für die reine und angewandte Mathematik 49:123–141

A. Seagar, *Application of Geometric Algebra to Electromagnetic Scattering*,
DOI 10.1007/978-981-10-0089-8

18. Grassmann H (1877) Der ort der Hamilton'schen quaternionen in der Ausdehnungslehre. Math Ann 12:375–386
19. Grassmann H (1995) On the various types of multiplication. A new branch of mathematics: the ausdehnungslehre of 1844, and other works. Open Court, Chicago, pp 451–468 (translated by Lloyd C. Kannenberg)
20. Grassmann H (1995) The position of the Hamiltonian quaternions in extension theory. A new branch of mathematics: the ausdehnungslehre of 1844, and other works. Open Court, Chicago, pp 525–538 (translated by Lloyd C. Kannenberg)
21. Einstein A (1905) Zur elektrodynamik bewegter körper. Annalen der Physik 322:891–921
22. Seagar A (2012) Calculation of electromagnetic fields in three dimensions using the Cauchy integral. IEEE Trans Mag 48(2):175–178
23. Kaczmarz S (1937) Angenäherte auflösung von systemen linearer gleichungen. Bulletin International de l'Académie Polonaise des Sciences et des Lettres. Classe des Sciences Mathématiques et Naturelles. Serie A. Sciences Mathématiques 35:355–357
24. Kaczmarz S (1993) Approximate solutions of systems of linear equations. Int J Control 57(6):1269–1271
25. Richardson LF (1911) The approximate arithmetical solution by finite differences of physical problems involving differential equations, with an application to the stresses in a masonry dam. Philos Trans R Soc Lond A 210(459–470):307–357
26. Richardson LF, Arthur Gaunt J (1927) The deferred approach to the limit. Philos Trans R Soc Lond A 226(636–646):299–361
27. Badea C, Grivaux S, Müller V (2010) The rate of convergence in the method of alternating projections. Arxiv.Org. arXiv:1006.2047v1 [math.FA]
28. Badea C, Lyubich Y (2010) Geometric, spectral and asymptotic properties of averaged products of projections in Banach spaces. Arxiv.Org. arXiv:1006.2052v2 [math.FA]
29. Ramo S, Whinnery JR, van Duzer T (1965) Fields and waves in communication electronics. Wiley, New York
30. Hestenes D (1993) Differential forms in geometric calculus. In: Brackx F (ed) Clifford algebras and their applications in mathematical physics. Fundamental theories of physics, vol 55. Kluwer, Dordrecht, pp 269–285
31. Ryan J, Struppa D (1998) Dirac operators in analysis. Pitman research notes in mathematics, vol 394. Longman, Harlow
32. Dirac PAM (1928) The quantum theory of the electron. Proc R Soc Ser A, Contain Pap Math Phys Character 117(778):610–624
33. Dirac PAM (1930) A theory of electrons and protons. Proc R Soc Ser A, Contain Pap Math Phys Character 126(801):360–365
34. McIntosh A, Mitrea M (1999) Clifford algebras and Maxwell's equations in Lipschitz domains. Math Methods Appl Sci 22(18):1599–1620
35. Pompeiu D (1912) Sur une class de fonctions d'une variable complexe. Rendiconti del Circolo Matematico di Palermo 33(1):108–113
36. Pompeiu D (1913) Sur une class de fonctions d'une variable complexe et sur certaines équations intégrales. Rendiconti del Circolo Matematico di Palermo 35(1):277–281
37. Mitrea M (1996) Hypercomplex variable techniques in harmonic analysis. In: Ryan John (ed) Clifford algebras in analysis and related topics. Studies in advanced mathematics, vol 21. CRC Press, Boca Raton, pp 103–128
38. Sprössig W (1998) On the treatment of classes of nonlinear boundary value and extension problems by Clifford analytic methods. In: Ryan John, Struppa Daniele (eds) Dirac operators in analysis. Pitman research notes in mathematics, vol 394. Longman, Harlow, pp 58–74
39. Chantaveerod A, Seagar AD, Angkaew T (2007) Calculation of electromagnetic field with integral equation based on Clifford algebra. Progress in electromagnetic research symposium (PIERS'07). Prague, Czech Republic, pp 62–71. 27–30 Aug 2007
40. Chantaveerod A (2009) Analysis and applications of multi-dimensional cauchy integral equation based on clifford number to scattering and radiation of electromagnetic fields. Ph.D. thesis, Chulalongkorn University, Bangkok, Thailand

41. Tsai LL, Dudley DG, Wilton DR (1974) Electromagnetic scattering by a three-dimensional rectangular conducting box. J Appl Phys 45(10):4393–4400
42. McGahan RV (1985) Scattering experiments at the Ipswich electromagnetic measurements facility: Backscatter from metal cubes. Technical Report RADC-TR-85-24, Rome Air Development Center, Griffiss Air Force Base, US-NY
43. Yaghjian AD, McGahan RV (1985) Broadside radar cross section of the perfectly conducting cube. IEEE Trans Antennas Propag 33(3):321–329
44. Horner WG (1819) A new method of solving numerical equations of all orders, by continuous approximation. Philos Trans R Soc Lond 109:1308–1335
45. Knuth DE (1969) The art of computer programming, seminumerical algortihms, vol 2. Addison-Wesley, Reading
46. Quarteroni A, Sacco R, Saleri F (2000) Numerical mathematics. Texts in Applied Mathematics, vol 37. Springer, New York
47. Wilkinson JH (1963) Rounding errors in algebraic processes., Notes on applied science No. 32Her Majesty's Stationary Office, London
48. Chantaveerod A, Seagar A (2009) Iterative solutions for electromagnetic fields at perfectly reflective and transmissive interfaces using Clifford algebra and the multi-dimensional Cauchy integral. IEEE Trans Antennas Propag 57(11):3489–3499
49. Badea C, Grivaux S, Müller V (2012) The rate of convergence in the method of alternating projections. St. Petersbg Math J 23(3):413–434
50. Coppersmith D, Winograd S (1982) On the asymptotic complexity of matrix multiplication. SIAM J Comput 11(3):472–492
51. Iliopoulos CS (1989) Worst-case complexity bounds on algorithms for computing the canonical structure of finite Abelian groups and the Hermite and Smith normal forms of an integer matrix. SIAM J Comput 18(4):658–669
52. Strassen V (1969) Gaussian elimination is not optimal. Numerische Mathematik 13(4):354–356
53. Harrington RF (1968) Field computation by moment methods. Macmillan, New York (reprinted John Wiley 1993)
54. Harrington RF (1987) The method of moments in electromagnetics. J Electromagn Waves Appl 1(3):181–200
55. Davidson DB (2010) Computational electromagnetics for RF and microwave engineering, 2nd edn. Cambridge University Press, Cambridge
56. Thomas R, Pär I, Anders B (2013) Computational electromagnetics. Texts in applied mathematics, vol 51. Springer, New York
57. Komech AI, Schubin MA (1999) Elements of the modern theory of partial differential equations. Springer, Berlin
58. Reutskiy SY (2006) The method of fundamental solutions for Helmholtz eigenvalue problems in simply and multiply connected domains. Eng Anal Bound Elem 30(3):150–159
59. Seagar A, Chantaveerod A (2015) An EM scattering algorithm for all materials. In: International conference on electromagnetics in advanced applications (ICEAA'15). Turin, Italy, p 256–259. 7–11 Sept 2015
60. Ergül Ö, Gürel L (2007) Fast and accurate solutions of scattering problems involving dielectric objects with moderate and low contrasts. In: Computational electromagnetics workshop, 2007. Izmir, Turkey, pp 59–64. 30–31 Aug 2007
61. Ergül Ö, Gürel L (2007) Accurate solutions of scattering problems involving low-contrast dielectric objects with surface integral equations. In: The second european conference on antennas and propagation (EuCAP 2007). Edinburgh, pp 1–6. 11–16 Nov 2007
62. Rao SM, Wilton DR, Glisson AW (1982) Electromagnetic scattering by surfaces of arbitrary shape. IEEE Trans Antennas Propag 30(3):409–418
63. Rao SM, Wilton DR (1990) E-field, H-field, and combined field solution for arbitrarily shaped three-dimensional dielectric bodies. Electromagnetics 10(4):407–421
64. Ylä-Oijala P, Taskinen M (2003) Calculation of CFIE impedance matrix elements with RWG and $n \times$ RWG functions. IEEE Trans Antennas Propag 51(8):1837–1846

65. Lloyd TW, Song JM, Yang M (2005) Numerical study of surface integral formulations for low-contrast objects. IEEE Antennas Wirel Propag Lett 4(1):482–485
66. Ylä-Oijala P (2008) Numerical analysis of combined field integral equation formulations for electromagnetic scattering by dielectric and composite objects. Prog Electromagn Res C 3:19–43
67. Seagar AD, Grognard RJ-M (1991) Simulation of current flow in piecewise constant media. Australas Phys Eng Sci Med 14(4):204–212
68. Auscher P, Hofmann S, Lacey M, Lewis J, McIntosh A, Tchamitchian P (2001) The solution of the Kato's conjectures. Comptes Rendus de l'Académie des Sciences 332(7):601–606
69. Auscher P, Hofmann S, Lacey M, McIntosh A, Tchamitchian P (2002) The solution of the Kato square root problem for second order elliptic operators on $\mathbb{R}^n$. Ann Math 156(2):633–654
70. Tchamitchian P (2001) The solution of Kato's conjecture (after Auscher, Hofmann, Lacey, McIntosh and Tchamitchian). Journées Équations aux Dérivées Partielles, pp 1–14
71. MatWeb LLC. MatWeb material property data (2015). http://www.matweb.com. Accessed 11 Feb 2015
72. Lide DR (1997) CRC handbook of chemistry and physics, 78th edn. CRC Press, Boca Raton
73. Aylward GH, Findlay TJV (1971) SI chemical data. Wiley, Sydney
74. Kinchin GH (1953) The electrical properties of graphite. Proc R Soc Lond Ser A, Math Phys Sci 217(1128):9–26
75. Grove AS (1967) Physics and technology of semiconductor devices. Wiley, New York
76. Washburn EW (1929) International critical tables of numerical data, physics, chemistry and technology, vol 6. McGraw Hill, New York
77. Kaye GWC, Laby TH (1995) Tables of physical and chemical constants, 16th edn. Longman Scientific & Technical, Essex
78. Jain H, Nowick AS (1982) Electrical conductivity of synthetic and natural quartz crystals. J Appl Phys 53(1):477–484

Index

A. Seagar, *Application of Geometric Algebra to Electromagnetic Scattering*,
DOI 10.1007/978-981-10-0089-8

MIX
Papier aus verantwortungsvollen Quellen
Paper from responsible sources
FSC® C105338

If you have any concerns about our products,
you can contact us on
ProductSafety@springernature.com

In case Publisher is established outside the EU,
the EU authorized representative is:
Springer Nature Customer Service Center GmbH
Europaplatz 3, 69115 Heidelberg, Germany

Printed by Libri Plureos GmbH
in Hamburg, Germany